Six-Minute Solutions
for Civil PE Exam Transportation Problems

Second Edition

Norman R. Voigt, PE, PLS

Professional Publications, Inc. • Belmont, CA

How to Locate and Report Errata for This Book

At Professional Publications, we do our best to bring you error-free books. But when errors do occur, we want to make sure you can view corrections and report any potential errors you find, so the errors cause as little confusion as possible.

A current list of known errata and other updates for this book is available on the PPI website at **www.ppi2pass.com/errata**. We update the errata page as often as necessary, so check in regularly. You will also find instructions for submitting suspected errata. We are grateful to every reader who takes the time to help us improve the quality of our books by pointing out an error.

SIX-MINUTE SOLUTIONS FOR CIVIL PE EXAM TRANSPORTATION PROBLEMS
Second Edition

Current printing of this edition: 1

Printing History

edition number	printing number	update
1	1	New book.
1	2	Minor corrections.
2	1	Code updates and minor corrections. Copyright update.

Copyright © 2006 by Professional Publications, Inc. All rights reserved. No part of this publication may be reproduced, stored in a retrieval system, or transmitted, in any form or by any means, electronic, mechanical, photocopying, recording, or otherwise, without the prior written permission of the publisher.

Printed in the United States of America

Professional Publications, Inc.
1250 Fifth Avenue, Belmont, CA 94002
(650) 593-9119
www.ppi2pass.com

Library of Congress Cataloging-in-Publication Data
Voigt, Norman R.
    Six-Minute solutions for civil PE exam transportation problems / Norman R. Voigt.
       p. cm.
    ISBN-13: 978-1-59126-052-3
    ISBN-10: 1-59126-052-3
    1. Transportation engineering--Examinations, questions, etc. I. Title.

TA1145.V65 2006
629.04076--dc22

                                                2005053502

Table of Contents

ABOUT THE AUTHOR iv

PREFACE AND ACKNOWLEDGMENTS v

INTRODUCTION

    Exam Format vii

    This Book's Organization vii

    How to Use This Book vii

REFERENCES ix

CODES USED TO PREPARE THIS BOOK xi

NOMENCLATURE xiii

BREADTH PROBLEMS

    Traffic Analysis 1

    Construction 2

    Geometric Design 3

DEPTH PROBLEMS

    Traffic Analysis 7

    Transportation Planning 12

    Construction 13

    Geometric Design 16

    Traffic Safety 20

BREADTH SOLUTIONS

    Traffic Analysis 23

    Construction 25

    Geometric Design 28

DEPTH SOLUTIONS

    Traffic Analysis 35

    Transportation Planning 49

    Construction 50

    Geometric Design 56

    Traffic Safety 64

About the Author

Norman R. Voigt, PE, PLS, is a registered civil engineer and land surveyor in Pennsylvania. Mr. Voigt obtained his bachelor's and master's degrees in civil engineering from the University of Pittsburgh. He has worked in both the private and public sectors, performing design, construction, maintenance, and inspection on a variety of highway and transit projects. For more than 25 years, he has taught sections of the civil PE review course given by Pennsylvania State University. He is currently on the adjunct faculty at the University of Pittsburgh.

Preface and Acknowledgments

The Principles and Practice of Engineering examination (PE exam) for civil engineering, prepared by the National Council of Examiners for Engineering and Surveying (NCEES), is developed from sample problems submitted by educators and professional engineers representing consulting, government, and industry. PE exams are designed to test examinees' understanding of both conceptual and practical engineering concepts. Problems from past exams are not available from NCEES or any other source. However, NCEES does identify the general subject areas covered on the exam.

The topics covered in *Six-Minute Solutions for Civil PE Exam Transportation Problems* coincide with those subject areas identified by NCEES for the transportation engineering depth module of the civil PE exam. Included among these problem topics are traffic analysis, transportation planning, construction, geometric design, and traffic safety.

The problems presented in this book are representative of the type and difficulty of problems you will encounter on the PE exam. These are both conceptual and practical, and they are written to provide varying levels of difficulty. Though you probably won't encounter problems on the exam exactly like those presented here, reviewing these problems and solutions will increase your familiarity with the exam problems' form, content, and solution methods. This preparation will help you considerably during the exam.

Problems and solutions have been carefully prepared and reviewed to ensure that they are appropriate and understandable, and that they were solved correctly. If you find errors or discover an alternative, more efficient way to solve a problem, please bring it to PPI's attention so your suggestions can be incorporated into future editions. You can report errors and keep up with the changes made to this book, as well changes to the exam, by logging on to Professional Publications' website at www.ppi2pass.com/errata.

I would like to thank Maria Redlinger for typing these problems for me. She took on the mystery of transforming my handwritten notes into presentable material. The staff at Professional Publications, especially Sarah Hubbard, and Heather Kinser, have been most helpful in encouraging me to complete the project, even when I thought I wasn't making any progress. I especially want to thank the technical reviewer, Dr. Maher Murad, for his invaluable comments.

Norman R. Voigt, PE, PLS

Introduction

EXAM FORMAT

The Principles and Practice of Engineering examination (PE exam) in civil engineering is an 8-hour exam divided into a morning and an afternoon session. The morning session is known as the "breadth" exam and the afternoon is known as the "depth" exam.

The morning session includes 40 problems from all of the five civil engineering subdisciplines (environmental, geotechnical, structural, transportation, and water resources), each subdiscipline representing about 20% of the problems. As the "breadth" designation implies, morning session problems are general in nature and wide-ranging in scope.

The afternoon session allows the examinee to select a "depth" exam module from one of the five subdisciplines. The 40 problems included in the afternoon session require more specialized knowledge than those in the morning session.

All problems from both the morning and afternoon sessions are multiple choice. They include a problem statement with all required defining information, followed by four logical choices. Only one of the four options is correct. Nearly every problem is completely independent of all others, so an incorrect choice on one problem typically will not carry over to subsequent problems.

Topics and the approximate distribution of problems on the afternoon session of the civil transportation exam are as follows.

Transportation: approx. 65% of exam problems
- Traffic Analysis
- Transportation Planning
- Construction
- Geometric Design
- Traffic Safety

Geotechnical: approx. 15% of exam problems
- Subsurface Exploration and Sampling
- Engineering Properties of Soils
- Soil Mechanics Analysis
- Shallow Foundations

Water Resources: approx. 20% of exam problems
- Hydraulics
- Hydrology

For further information and tips on how to prepare for the civil transportation engineering PE exam, consult the *Civil Engineering Reference Manual* or Professional Publications' website, www.ppi2pass.com.

THIS BOOK'S ORGANIZATION

Six-Minute Solutions for Civil PE Exam Transportation Problems is organized into two sections. The first section, Breadth Problems, presents 20 transportation engineering problems of the type that would be expected in the morning part of the civil engineering PE exam. The second section, Depth Problems, presents 80 problems representative of the afternoon part of this exam. The two sections of the book are further subdivided into the topic areas covered by the transportation exam.

Most of the problems are quantitative, requiring calculations to arrive at a correct solution. A few are non-quantitative. Some problems will require a little more than 6 minutes to answer and others a little less. On average, you should expect to complete 80 problems in 480 minutes (8 hours), or spend 6 minutes per problem.

Six-Minute Solutions for Civil PE Exam Transportation Problems does not include problems related directly to geotechnical and water resources engineering, although problems from these subdisciplines will be included in the transportation exam. *Six-Minute Solutions for Civil PE Exam Geotechnical Problems* and *Six-Minute Solutions for Civil PE Exam Problems: Water Resources* provide problems for review in these areas of civil engineering.

HOW TO USE THIS BOOK

In *Six-Minute Solutions for Civil PE Exam Transportation Problems*, each problem statement, with its supporting information and answer choices, is presented in the same format as the problems encountered on the PE exam. The solutions are presented in a step-by-step sequence to help you follow the logical development of the correct solution and to provide examples of how you may want to approach your solutions as you take the PE exam.

Each problem includes a hint to provide direction in solving the problem. In addition to the correct solution, you will find an explanation of the faulty solutions leading to the three incorrect answer choices. The incorrect solutions are intended to represent common mistakes made when solving each type of problem. These may be simple mathematical errors, such as failing to square a term in an equation, or more serious errors, such as using the wrong equation.

To optimize your study time and obtain the maximum benefit from the practice problems, consider the following suggestions.

1. Complete an overall review of the problems and identify the subjects that you are least familiar with. Work a few of these problems to assess your general understanding of the subjects and to identify your strengths and weaknesses.

2. Locate and organize relevant resource materials. (See the references section of this book as a starting point.) As you work problems, some of these resources will emerge as more useful to you than others. These are what you will want to have on hand when taking the PE exam.

3. Work the problems in one subject area at a time, starting with the subject areas that you have the most difficulty with.

4. When possible, work problems without using the hint. Always attempt your own solution before looking at the solutions provided in the book. Use the solutions to check your work or to provide guidance in finding solutions to the more difficult problems. Use the incorrect solutions to help identify pitfalls and to develop strategies to avoid them.

5. Use each subject area's solutions as a guide to understanding general problem-solving approaches. Although problems identical to those presented in *Six-Minute Solutions for Civil PE Exam Transportation Problems* will not be encountered on the PE exam, the approach to solving problems will be the same.

Solutions presented for each example problem may represent only one of several methods for obtaining a correct answer. Although we have tried to prepare problems with unique solutions, alternative problem-solving methods may produce a different, but nonetheless appropriate, answer.

References

The minimum recommended library for the civil exam consists of PPI's *Civil Engineering Reference Manual*. You may also find the following references helpful in completing some of the problems in *Six-Minute Solutions for Civil PE Exam Transportation Problems*.

American Association of State Highway and Transportation Officials (AASHTO). *A Policy on Geometric Design of Highways and Streets* (GDHS). Washington, DC.

American Concrete Institute. *Building Code Requirements for Structural Concrete and Commentary*, (ACI 318/318R). Farmington Hills, MI.

Asphalt Institute. *The Asphalt Handbook*, Manual Series No. 4 (MS-4). College Park, MD.

_____. *Superpave Mix Design*, Manual Series No. 2 (SP-2). College Park, MD.

Banks, James H. *Introduction to Transportation Engineering*. New York: McGraw-Hill.

Brinker, Russell C., and Roy Minnick, eds. *The Surveying Handbook*. New York: Van Nostrand Reinhold Co.

Chrest, Anthony P., Mary S. Smith, and Sam Bhuyan. *Parking Structures: Planning, Design, Construction, Maintenance and Repair*. New York: Van Nostrand Reinhold.

Federal Highway Administration. *Manual on Uniform Traffic Control Devices for Streets and Highways* (MUTCD). Washington, DC.

Garber, Nicholas J., and Lester A. Hoel. *Traffic and Highway Engineering*. Boston, MA: PWS Publishing Company.

Hickerson, Thomas F. *Route Location and Design*. New York: McGraw-Hill.

Homburger, Wolfgang S., et al. *Fundamentals of Traffic Engineering*. Berkeley, CA: Institute of Transportation Studies, University of California.

Horonjeff, Robert, and Francis X. McKelvy. *Planning and Design of Airports*. New York: McGraw-Hill.

Institute of Transportation Engineers (ITE). *Manual of Traffic Signal Design*. New Jersey: Prentice Hall.

_____. *Manual of Transportation Engineering Studies*. Washington, DC.

_____. *Traffic Engineering Handbook*. New Jersey: Prentice Hall.

Khisty, C. Jotin, and B. Kent Lall. *Transportation Engineering: An Introduction*. New Jersey: Prentice Hall.

Lindeburg, Michael R. *Civil Engineering Reference Manual for the PE Exam*. Belmont, CA: Professional Publications, Inc.

Mannering, Fred L., and Walter P. Kilareski. *Principles of Highway Engineering and Traffic Analysis*. New York: John Wiley & Sons.

Merritt, Frederick S., M. Kent Loftin, and Jonathan T. Ricketts. *Standard Handbook for Civil Engineers*. New York: McGraw-Hill.

Papacostas, C. S., and P. D. Prevedouros. *Transportation Engineering and Planning*. New Jersey: Prentice-Hall.

Portland Cement Association (PCA). *Design and Control of Concrete Mixtures*. Skokie, IL.

Transportation Research Board. *Highway Capacity Manual*, Special Report 209 (HCM). Washington, DC.

Wright, Paul H., and Karen K. Dixon. *Highway Engineering*. New York: John Wiley & Sons.

Codes Used to Prepare This Book

At the time of printing, the codes, policies, handbooks, and design guides used to write these problems corresponded with those announced by the NCEES as being the basis of the transportation portion of the Civil PE Exam. However, the codes used on the exam are subject to change as the NCEES sees fit. For a current list of the codes used by the NCEES, refer to PPI's website (www.ppi2pass.com).

American Association of State Highway and Transportation Officials (AASHTO). *A Policy on Geometric Design of Highways and Streets* (GDHS). Washington D.C., 2001.

Asphalt Institute. *The Asphalt Handbook*, Manual Series No. 4 (MS-4). College Park, MD, 1989.

Federal Highway Administration. *Manual on Uniform Traffic Control Devices for Streets and Highways* (MUTCD). Washington D.C., 2003.

Institute of Transportation Engineers (ITE). *Traffic Engineering Handbook*, 5th ed. New Jersey: Prentice Hall, 1999.

Portland Cement Association (PCA). *Design and Control of Concrete Mixtures*, 14th ed. Skokie IL, 2002.

Transportation Research Board. *Highway Capacity Manual*, Special Report 209 (HCM). Washington D.C., 2000.

Nomenclature

a	acceleration	ft/sec^2, m/s^2	i	annual growth rate	decimal
A	accident	veh-mi	I	angle of intersection between two tangents	deg
A	area	ft^2	I	deflection of a curve	deg
A	number of autos per household	–	I_w	width of parking aisle	ft
A	top dimension	ft	$I_{w,\text{han}}$	width of handicapped stall aisle	ft
A	total grade change in vertical curve	%	K	probability factor determined by desired level of significance	–
ADT	average daily traffic	veh/day			
B	base dimension	ft	K	ratio of vertical curve length to grade change	ft/%
BF	bulking factor	decimal			
BFFS	base free-flow speed	mph	L	length	ft
c	cost of roadway excavation	$/yd^3	L	length of car	ft
C	capacity of roadway approach	vph	L	length of vertical curve	ft, sta
C	signal cycle length	sec	L	lost time per signal cycle	sec
CS	critical sum of flow ratios	–	LEF	load equivalent factor	–
D	degree of curve	deg	leh	length of economical haul	ft
D	density of flow	pcpmpl	lfh	length of freehaul	ft
D	depth of diagonal parking space	ft	loh	economical length of overhaul	ft
D	distance	ft	LP	low point of vertical curve	–
e	superelevation rate	decimal, %	L_s	length of spiral	ft
E	elevation	ft	LVC	length of vertical curve	sta
E	external distance	ft	M	mass	lbm
E	passenger car equivalent	–	M	middle ordinate of curve	ft
EAL	equivalent axle loading	–	N	number of maneuvers per hour	decimal
f	friction factor	–	O	center of circle	–
f	friction coefficient	–	O'	shifted center of circle	–
f	volume adjustment factor	decimal	O	vehicle occupancy	–
F	fuel consumption	gal/veh-hr	p	offset distance from shifted PC to spiral tangent	ft
F	future worth	$			
F	speed adjustment	mph	P	applied load	kips
FFS	free-flow speed	mph	P	number of persons per household	–
g	gravitational acceleration	ft/sec^2, m/s^2			
G	grade	decimal, ft/sta, %	P	persons	–
			P	present worth	$
GF	growth factor	–	P	proportion of traffic in a certain class	decimal
h	altitude of triangular shape	ft			
h	cost of overhaul	$/yd^3-sta	PHF	peak hour factor	–
h	depth of aggregate layer	in	R	accident rate at a location	–
h	depth of cut	ft	R	curve radius	ft
h	elapsed time between bicycle arrivals	min/bicycle	R	rate of change in grade	%/sta
			s	distance	ft, m
H	number of handicapped parking spaces	–	s	saturation flow rate for lane group	vph
H	height	ft			

S	salvage value	$	C	cut	
S	sight distance	ft	crit	critical	
s_0	base saturation flow rate	vph	decel	deceleration	
S_d	depth of parking stall 90° to aisle	ft	E	effective	
S_l	length of parking stall point of reverse spiral	ft –	f F	following fill	
S_w	width of parking stall	ft	g	grade or lane group	
$S_{w,\text{han}}$	width of handicapped parking stall	ft	gl	lane in the lane group with the highest volume	
t	time of queue dissipation	min	han	handicapped	
t	travel time	sec	HV	heavy vehicle	
T	number of person-trips per household per day	–	horiz i	horizontal direction attribute type or ith approach	
T	tangent length	ft	LC	lateral clearance	
T_A	number of auto trips per day	–	Lpb	pedestrian-blockage for left turns	
TF	truck factor	–	LT	left turn	
v	velocity	ft/sec, mph, kph	LU	lane utilization	
V	flow rate	vph, pers/hr	LW m	lane width median type or parking maneuvers per hour	
V	volume	ft³, yd³	M	median	
V_g	unadjusted demand flow rate	vph	max	maximum	
VM	vehicle miles traveled	veh-mi	min	minimum	
V_p	rate of flow during peak period	pcphpl	n o	normal initial or office use	
w	weight	lbf	O	optimal	
w	width	ft	p	driver population, parking, or peak	
W	walkway width	ft	prop	proposed	
x	horizontal distance from PVC	ft	r	perception-reaction time or retail use	
x	tangent distance	ft	R	recreational vehicle	
X	degree of saturation for signal phase	decimal	Rpb RT	pedestrian-blockage for right turns right turn	
X	horizontal distance from PVC	ft	s	spiral or stopping	
y	offset from tangent	ft	seg	segment	
Z	distance	ft	sov sp T	single occupant vehicle parking space trucks and buses	

Symbols

			vert	vertical	
α	angle	deg	w	width	
β	angle	deg	w/s	warehouse/storage use	
δ	difference	–			
Δ	deflection angle	deg			
η	efficiency	decimal			
μ	mean service rate per server	hr^{-1}			

Subscripts

0	initial
15	peak 15 minutes
a	area
A	number of access points per mile
accel	acceleration
ave	average
b	base dimension or braking
B	base width or bus maneuvers
bb	bus blockage
c	carpool

Breadth Problems

TRAFFIC ANALYSIS

PROBLEM 1

The average number of cars passing a point is 2200 pcphpl. The cars travel at an average speed of 42 mph. If the average length of a car is 19 ft, the distance between the cars is most nearly

- (A) 43 ft
- (B) 82 ft
- (C) 100 ft
- (D) 130 ft

Hint: Look for a common element among the defining units, which are related to speed, time, distance, and the number of cars.

PROBLEM 2

A freeway in rolling terrain has the following characteristics.

commuter traffic volume (one way)	1970 vph
number of lanes (in each direction)	4
percentage of trucks	3%
percentage of buses	3%
percentage of RVs	1%
peak hour factor (PHF)	0.85

What is most nearly the peak hour flow rate?

- (A) 460 pcphpl
- (B) 580 pcphpl
- (C) 600 pcphpl
- (D) 640 pcphpl

Hint: The peak hour flow rate is the per-lane passenger-car equivalent of the hourly count of total vehicle flow. The formulas for equivalent passenger-car flow rates from the HCM must be used.

PROBLEM 3

A 10 mi section of freeway in mountainous terrain has the characteristics listed. 90% of the traffic is made up of local commuters, and the remainder is a mixture of through traffic and recreational traffic to a nearby national park.

design speed	60 mph
number of lanes (in each direction)	3
lane width	11 ft
left shoulder width	4 ft
right shoulder width	6 ft
percentage of trucks	3%
percentage of buses	2%
percentage of RVs	3%
number of interchanges	4
peak hour factor (PHF)	0.90
driver population factor	0.95
hourly traffic volume	4495 vph

What is the level of service (LOS) of this section?

- (A) C
- (B) D
- (C) E
- (D) F

Hint: Heavy vehicle factors for RVs are not always included with trucks and buses.

PROBLEM 4

A four-lane divided highway in a suburban area has the following characteristics.

lane width	10 ft
average grade	<2%
left clearance	4 ft
right clearance	2 ft
percentage of heavy vehicles	7%
access spacing	300 ft
design speed	60 mph
directional design hour volume	2540 vph

What is the level of service (LOS) of the highway?

- (A) C
- (B) D
- (C) E
- (D) F

Hint: Roadway configuration restrictions affect the free-flow speed (FFS). Design speed can be considered to be the base free-flow speed (BFFS) if there is no information to the contrary.

PROBLEM 5

On a highway facility, how does the observed hourly vehicle volume differ from the design peak-period flow rate?

(A) The observed hourly vehicle volume is divided by the number of hours covered by the observation to obtain the peak-period flow rate.

(B) The highest 15 min vehicle volume is divided by the highest 1 hr flow rate to obtain the peak-period flow rate.

(C) The observed hourly vehicle volume includes a mix of heavy vehicles, while the design peak-period flow rate has been adjusted for passenger-car equivalents of heavy vehicles, the peak hour factor, the driver population, and the number of lanes.

(D) The observed vehicle volume is divided by the number of hours covered by the observation and the number of lanes over which the observation took place.

Hint: Various vehicle sizes must be converted to a common unit of vehicle measure.

CONSTRUCTION

PROBLEM 6

A highway project requires 30,000 yd^3 of embankment (fill). There are 24,000 yd^3 of earth loam cut on site. The earth loam has a shrinkage factor of 10%. A borrow pit is available that has sandy loam with an in-place mass of 120 lbm/ft^3. The sandy loam has a swell factor of 18% and a shrinkage factor of 10%. Trucks available to haul the borrow material can carry 48,000 lbm per load with a 15 yd^3 maximum capacity. How many truckloads of borrow material are required?

(A) 472
(B) 498
(C) 624
(D) 734

Hint: When soil is excavated for hauling, it increases in unit volume, or bulks. After it is placed in a compacted fill, the density increases from its natural state, causing the soil to shrink.

PROBLEM 7

A highway project requires 6500 yd^3 of sand, which is available from three sources, each at a different hauling distance from the site.

source	cost per ton ($)	source in-place specific weight (lbf/ft^3)	haul distance (mi)	hauling cost ($/ton-mi)	bulking factor (%)
I	5.40	130	20	0.23	11
II	5.70	130	17	0.23	17
III	8.10	130	12	0.23	12

The sand will be placed at 98% compaction. When delivered to the site, the sand that will cost the least per cubic yard will be from

(A) source I
(B) source II
(C) source III
(D) source I or II

Hint: Since the density and hauling cost per ton are the same for all three sources, the main variation in delivered cost to the site will originate from the source cost and the haul distance.

PROBLEM 8

Following is a list of tests performed on highway paving material and the purposes of the tests.

test	type	use (in highway pavements)
I	specific gravity	convert weights to volumes
II	solubility	determine asphalt content
III	viscosity	grading
IV	air entrainment	frost susceptibility
V	ductility	brittleness (check for poor performance)
VI	temperature susceptibility	performance grading
VII	water content	strength
VIII	flash point	hazard rating
IX	surface tension	miscibility
X	distillation	grading

Which of the tests have NO USE for asphalt material?

(A) IV, VIII
(B) III, IV, X
(C) IV, VII, IX
(D) IV, V, VII, VIII

Hint: Many tests are related to the unique properties of either concrete or asphalt materials, but not both. Asphalt does not use water in the mix and is made from byproducts of petroleum. Liquid asphalt is delivered at high temperature as an emulsion to be mixed with aggregate, which must be dry.

PROBLEM 9

Project activities are listed showing activity predecessors, successors, and time durations. The project start is at 0 days.

activity	predecessors	successors	duration (days)
start	–	A	0
A	start	B,D,E,F	7
B	A	C	7
C	B	D,I	1
D	A,C	G,I	5
E	A	G,H	14
F	A	H,J	6
G	D,E	J	17
H	E,F	J	11
I	C,D	J,finish	8
J	F,G,H,I	finish	6
finish	J,I	–	0

What is the project duration?

(A) 26 days
(B) 29 days
(C) 44 days
(D) 52 days

Hint: To find the critical path, look for the activities with the longest time durations, especially those that have multiple predecessors and/or multiple successors.

PROBLEM 10

In the critical path method (CPM), each *critical activity* has the following characteristics.

1. The early start (ES) and late start (LS) times of the activity must be equal.

2. The early finish (EF) and late finish (LF) times of the activity must be equal.

What is a third characteristic?

(A) Float occurs between the early and late event times at the finish.
(B) The longest-duration activities determine the critical path.
(C) Slack time along the critical path equates to the delay in the project that can be permitted without causing the project to fall behind schedule.
(D) The difference between ES and LF must equal the duration of the event.

Hint: Each activity along the critical path starts as soon as the last previous activity is finished.

GEOMETRIC DESIGN

PROBLEM 11

A circular horizontal curve with a radius of 1430 ft and a deflection angle of 14°30′ Rt has its point of intersection (PI) at sta 572+00. The forward tangent is to be shifted parallel to itself and 12 ft to the left. If the point of curvature (PC) station remains in the same place, what is most nearly the radius of the shifted curve?

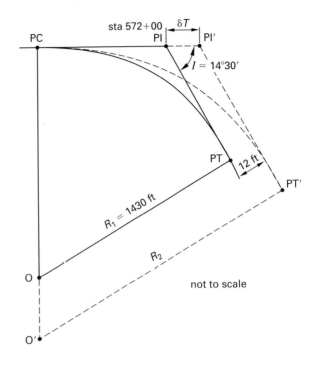

(A) 1050 ft
(B) 1530 ft
(C) 1810 ft
(D) 1820 ft

Hint: The curve deflection remains the same for the shifted curve. A new tangent length is increased by extending the back tangent to intersect the shifted position of the ahead tangent.

PROBLEM 12

Referring to the following illustration, subtangent AB is 330 ft, angle α is 25°, and angle β is 38°. What is most nearly the radius of the curve that will be tangent to lines CI, AB, and IE?

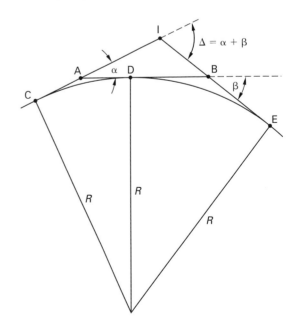

(A) 270 ft
(B) 320 ft
(C) 580 ft
(D) 4300 ft

Hint: The angles given are the deflections of each of the curve arcs, which can be treated as two independent curves with the same radius.

PROBLEM 13

Referring to the following illustration, AG is 200 ft, angle α is 90°, and the degree of curve is 2°. What is most nearly the distance GP?

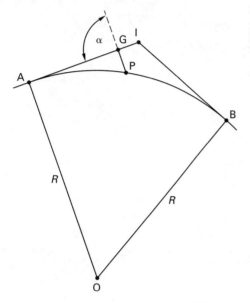

(A) 1.8 ft
(B) 3.5 ft
(C) 7.0 ft
(D) 28 ft

Hint: Look for a way to construct a triangle using point P as one corner.

PROBLEM 14

A horizontal curve has a deflection angle of 67°45′ at PI sta 126+50. When the external is close to 35 ft, what is most nearly the radius of the curve to the nearest whole degree of curve?

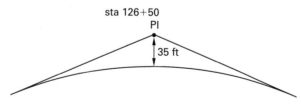

(A) 44 ft
(B) 170 ft
(C) 180 ft
(D) 210 ft

Hint: Solving initially for the radius then converting to degree of curve reduces the number of steps and the chance of errors.

PROBLEM 15

A circular horizontal curve has a radius of 2500 ft and a deflection angle of 35°. The point of intersection (PI) of the curve is at sta 485+26.75. What is the station of the point of tangent (PT)?

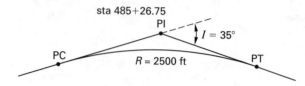

(A) sta 483+00
(B) sta 492+40
(C) sta 492+65
(D) sta 493+15

Hint: First, find the station of the point of curvature (PC) based on the tangent length of the curve.

PROBLEM 16

A curve on a low-speed urban street has a superelevation of −4%. The design is to follow GDHS for a design speed of 30 mph. What is most nearly the minimum radius for the curve?

(A) 16 ft
(B) 230 ft
(C) 300 ft
(D) 330 ft

Hint: A reverse cross-slope increases the feeling of discomfort when driving a curve at a particular speed.

PROBLEM 17

A −4% grade meets a +5% grade at sta 34+00. Using a 600 ft vertical curve, find the position of the low point (LP).

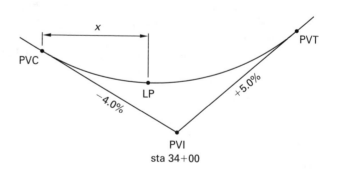

(A) 0.44 ft ahead from the PVC
(B) 2.66 ft ahead from the PVC
(C) 44.44 ft ahead from the PVC
(D) 266.70 ft ahead from the PVC

Hint: The low point is where the grade is equal to zero.

PROBLEM 18

A 500 ft long sag vertical curve passes under a bridge at sta 82+45. The point of vertical curve (PVC) is at sta 81+00. A −3.6% curve meets a +4.4% curve at the point of vertical intersection (PVI), which is at elevation 425.38 ft. What is most nearly the elevation of the point under the bridge?

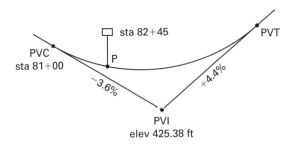

(A) 413 ft
(B) 430 ft
(C) 431 ft
(D) 441 ft

Hint: The elevation of any point on a vertical curve can be found as long as one other elevation point can be determined on the curve.

PROBLEM 19

A vertical curve is to be placed through sta 81+50 at an elevation of 638.42 ft. The grade into the curve is −3.6% and the grade out of the curve is +3.8%. The point of vertical curve (PVC) is at sta 76+00 and elevation 646.12 ft. What length of curve is required?

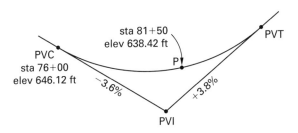

(A) 25.0 ft
(B) 168 ft
(C) 407 ft
(D) 925 ft

Hint: Curve elevations on parabolic vertical curves are found by comparison with other known elevation points. The relationship can be worked inversely to determine curve length if two points are already known.

PROBLEM 20

A specification for a high-speed train states that vertical acceleration in a parabolic sag vertical curve is to be limited to 0.03g. Using a design speed of 150 mph, a −0.75% grade meets a +0.45% grade. What is most nearly the minimum length of vertical curve required?

(A) 150 ft
(B) 280 ft
(C) 600 ft
(D) 1980 ft

Hint: The change in vertical velocity is related to time and travel speed when calculating the distance necessary to change vertical direction.

Depth Problems

TRAFFIC ANALYSIS

PROBLEM 21

An exclusive bikeway is operating with an average two-way flow of 20 bicycles in the peak 15 min period. The directional distribution is 60/40. What is the heaviest one-way density of flow at an average speed of 11.2 mph?

(A) 0.3 bicycles/mi
(B) 4.3 bicycles/mi
(C) 7.2 bicycles/mi
(D) 27 bicycles/mi

Hint: Density is related to the number of vehicles per unit length of the pathway.

PROBLEM 22

In order to specify traffic density as determined by the HCM, which of the following criteria are used?

I. passenger-car equivalents
II. running speed
III. vehicle count per unit of time
IV. vehicle length
V. vehicle weight
VI. vehicle spacing

(A) I, IV, and VI
(B) III, IV, and VI
(C) III, V, and VI
(D) I, II, IV, and VI

Hint: Traffic density is expressed as the number of passenger cars per mile, per lane of roadway.

PROBLEM 23

Determining service flow rate of a roadway requires which of the following criteria, according to the HCM?

I. driver population factor
II. heavy vehicle factor
III. lane width factor
IV. number of lanes
V. parking adjustment factor
VI. peak-hour factor
VII. average control delay per vehicle
VIII. volume-over-capacity ratio

(A) I, II, IV, and VI
(B) I, II, IV, and VIII
(C) I, II, III, IV, and V
(D) II, III, V, VI, and VII

Hint: Service flow rate is expressed as the number of passenger cars per hour, per lane of roadway.

PROBLEM 24

On freeways, according to the HCM, which of the following statements is TRUE?

(A) Trucks are usually a higher percentage of peak hour traffic than of off-peak traffic.
(B) On upgrades, higher percentages of trucks tend to have lower passenger-car equivalents.
(C) On steep downgrades, trucks have lower passenger-car equivalents than they do on slight downgrades.
(D) When the percentage of trucks and the percentage of RVs are nearly equal, the RVs can be combined with trucks and considered as one heavy vehicle factor.

Hint: A grouping of trucks on freeways tends to form a platoon, and the trucks tend to operate more uniformly than do passenger cars.

PROBLEM 25

As used in the HCM, the peak hour factor is determined by

(A) the peak hour flow divided by the peak 15 min flow rate, expressed as an hourly rate
(B) the peak hour flow divided by the maximum rate of flow that occurs in the peak interval, usually 5 min
(C) the peak interval of flow, usually 5 min, divided by the peak one-hour flow
(D) the peak 15 min of flow divided by the peak one-hour flow

Hint: The traffic flow rate varies throughout a 1 hr period and usually requires more than 5 min to adjust to a new rate.

8 SIX-MINUTE SOLUTIONS FOR CIVIL PE EXAM TRANSPORTATION PROBLEMS

PROBLEM 26

When describing highway traffic flow, which of the following statements is NOT true?

(A) A highway with buses in the traffic stream has a higher capacity of persons per hour than does a highway with only automobiles in the traffic stream.

(B) A highway with trucks in the traffic stream has a lower capacity of persons per hour than does a highway with only automobiles in the traffic stream.

(C) For a highway with a large number of trucks in the trafic stream (as compared to RVs), the RVs can be combined with the trucks when determining passenger-car equivalents of heavy vehicles.

(D) For a highway in mountainous terrain with long grades, the same passenger-car equivalent is used for trucks on upgrades as is used for trucks on downgrades.

Hint: Buses and passenger vans increase the person capacity of a highway, while trucks and RVs decrease the person capacity.

PROBLEM 27

Two lanes of a freeway have a capacity of 4200 vph. The normal flow on these two lanes is 3100 vph. An incident blocks both lanes for 10 min. After 10 min, both lanes are opened to full traffic flow. Approximately how long does it take to dissipate the queue that resulted from the blockage?

(A) 7.0 min
(B) 28 min
(C) 38 min
(D) 67 min

Hint: Departure from the blockage site is at full capacity, while arrival continues at the same rate as before the blockage.

PROBLEM 28

A four-lane freeway has a directional capacity of 2100 vph for each lane. The normal directional flow is 3100 vph. An incident blocks one lane for 20 min and then is cleared to allow the full traffic capacity flow. Approximately how long does it take to dissipate the queue after the blockage has been cleared?

(A) 4.8 min
(B) 30 min
(C) 38 min
(D) 57 min

Hint: Departure from the blockage is at two rates, while arrival continues at the same rate as before the blockage.

PROBLEM 29

The stopping distance for a car traveling at 50 mph is 461 ft, including a 2.5 sec perception-reaction time. If a car traveling at 60 mph is to stop in the same distance with the same friction factor, what is most nearly the required perception-reaction time?

(A) 0 sec
(B) 0.1 sec
(C) 0.7 sec
(D) 2 sec

Hint: The stopping distance includes braking distance and perception-reaction distance.

PROBLEM 30

A car traveling at 50 mph is followed by a car traveling at 60 mph. The lead car suddenly brakes to a stop within a 278 ft distance. Both cars have the same stopping friction factor, and the following car has a perception-reaction time of 2.0 sec. Assume the road is level; that is, the grade is 0 ft/ft.

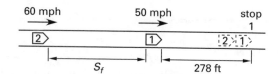

In order to avoid a collision, the minimum distance between the two cars must be most nearly

(A) 86 ft
(B) 280 ft
(C) 300 ft
(D) 320 ft

Hint: The following car is aware that the lead car is stopping when the driver sees the brake lights turn on.

PROBLEM 31

Passenger stations are spaced 1 mi apart on a rapid transit line. Trains accelerate out of a station at 5.5 ft/sec^2 and decelerate into a station at 4.4 ft/sec^2. Trains arrive every 5 min and have a top speed of 80 mph. What is most nearly the average speed of a train between stations?

(A) 12 mph
(B) 40 mph
(C) 52 mph
(D) 54 mph

Hint: The trip stages between stations consist of accelerating, traveling at constant speed, and decelerating.

PROBLEM 32

A rapid transit line has trains scheduled to arrive at a station every 5 min. The trains each have 4 cars and can carry 220 passengers per car. The trains have a 1 min dwell time at each station. What is most nearly the capacity of the transit line?

(A) 880 passengers/hr
(B) 2640 passengers/hr
(C) 10,600 passengers/hr
(D) 13,200 passengers/hr

Hint: The arrival rate and departure rate are the same, regardless of the length of dwell time at the station.

PROBLEM 33

A high-speed train is to reach 150 mph between stations. Acceleration is limited to $0.18g$ and deceleration is limited to $0.12g$. What is most nearly the minimum spacing between stations?

(A) 0.80 mi
(B) 0.90 mi
(C) 2.0 mi
(D) 6.5 mi

Hint: The train reaches 150 mph for an instant between stations.

PROBLEM 34

A commuter train has stations spaced 1 mi apart. The top speed of the train is 80 mph. The train accelerates at 5.5 ft/sec² and decelerates at 4.5 ft/sec². It dwells at the platform for 20 sec to pick up passengers. What is most nearly the average running speed of the train?

(A) 33 mph
(B) 41 mph
(C) 52 mph
(D) 64 mph

Hint: Running speed applies to the time during which the train is in motion between stations.

PROBLEM 35

A light-rail line 10 mi long currently makes an average of 4 stops/mi. Due to complaints of slow service, the number of stops is being reduced to 3 stops/mi. The vehicles accelerate at 3.5 mph/sec and decelerate at 2.5 mph/sec. The normal constant running speed is 35 mph. Each stop increases travel time by 30 sec, which includes decelerate, dwell, and accelerate times. Considering the reduced number of stops, what is most nearly the increase in average light-rail vehicle speed?

(A) 0.50 mph
(B) 2.5 mph
(C) 16 mph
(D) 19 mph

Hint: The average speed with no stops is the same as the constant running speed of 35 mph.

PROBLEM 36

Bus service is being planned for an arena event. There will be 12,000 people in attendance, 35% of whom will depart by bus. The buses have an average capacity of 62 persons/bus. The entire arena will empty in one hour at the end of the event. The peak hour factor is projected at 0.85. Approximately how many buses will be required during the highest peak 15 min after the event?

(A) 15
(B) 20
(C) 70
(D) 80

Hint: Here, peak hour factor is defined similarly to the highway traffic definition of peak hour factor.

PROBLEM 37

Transit planners are attempting to determine expected transit usage from a community that produces 12,000 trips/day. The population density is 10,000 persons/mi² and there is an average of 0.80 autos/household. The following model for the urban travel factor (UTF) has been established for the community.

$$\text{UTF} = \left(\frac{1}{1000}\right)\left(\frac{\text{households}}{\text{auto}}\right) \times \left(\frac{\text{persons}}{\text{mi}^2}\right)$$

$$\% \text{ transit usage} = \frac{\text{UTF}}{0.6}$$

Approximately how many residents can be expected to use transit?

(A) 1500
(B) 1600
(C) 2100
(D) 2500

Hint: Households per auto is the inverse of autos per household.

PROBLEM 38

The numbers for airport runways are assigned

(A) based on the heading bearing, rounded to the nearest 10°
(B) based on the airport design firm's internal design procedure
(C) according to the take-off heading, rounded to the nearest 10° and dropping the last zero
(D) according to the landing heading, or azimuth, rounded to the nearest 10° and dropping the last zero

Hint: The most critical condition is to find a place and direction to land, even under power failure and poor visibility.

PROBLEM 39

A transit station serves a stadium with a capacity of 50,000 persons. When an event is finished, 95% of the people in attendance are expected to leave in one hour, and the transit system is expected to carry 35% of those in attendance. The peak hour factor, defined similarly to the highway traffic definition of peak hour factor, is 0.75, and the level of service (LOS) is to be D. What is the minimum required effective width of the concourse walkway leading to the station?

(A) 25 ft
(B) 34 ft
(C) 76 ft
(D) 96 ft

Hint: Use platoon-adjusted LOS criteria.

PROBLEM 40

A paved parking area is being designed with 90° spaces that are 8.5 ft wide and 18 ft deep. The aisle is to be two-way and 24 ft wide. Drainage and maneuvering areas require 5% of the paved area.

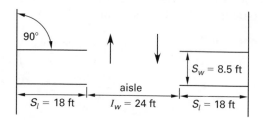

Approximately how many spaces can be provided per acre of pavement?

(A) 81 spaces/ac
(B) 120 spaces/ac
(C) 160 spaces/ac
(D) 170 spaces/ac

Hint: The area required for one space includes an adjoining section of aisle shared by another space.

PROBLEM 41

A parking lot is being designed for a capacity of 550 spaces. Spaces are to be 9 ft wide by 18 ft deep, placed at a 90° angle from a 22 ft wide aisle. Maneuvering areas and access driveways occupy 3% of the parking area.

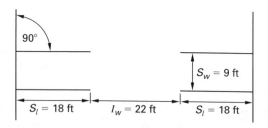

Federal regulations require that 2% of the total spaces be reserved for vehicles of disabled persons. Two of the handicapped spaces must accommodate van parking.

The handicapped spaces must be 8 ft wide with an adjoining 5 ft wide access aisle.

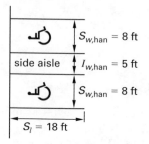

The handicapped van spaces must have an 8 ft access aisle. One access aisle may serve two abutting spaces.

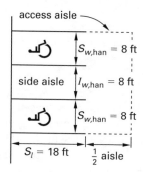

What is the approximate area required for the parking lot?

(A) 144,000 ft²
(B) 149,000 ft²
(C) 152,000 ft²
(D) 208,000 ft²

Hint: Assign one-half of the adjacent aisle width to the area required for each space.

PROBLEM 42

A 1.5 ac parking lot is to be layed out using the following configuration.

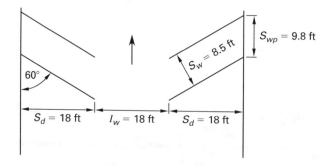

Driveways and maneuvering space require 10% of the total lot area. How many spaces will the lot accommodate?

(A) 110 spaces
(B) 222 spaces
(C) 247 spaces
(D) 256 spaces

Hint: A row module has the area of two spaces and the shared portion of the aisle between the spaces.

PROBLEM 43

An 80,000 ft² building is to be constructed. It will contain 45,000 ft² of daytime office space and 12,000 ft² of evening retail space. 10% of the total area will be assigned to mechanical space, and the remaining space will be for warehouse and storage use. The parking requirements are described by a certain zoning code as follows.

building space use	parking spaces required
offices	1 space/300 ft²
retail	1 space/150 ft²
warehouse/storage	1 space/500 ft²

Approximately how many parking spaces are required?

(A) 150 spaces
(B) 230 spaces
(C) 260 spaces
(D) 276 spaces

Hint: Each building space can be assigned only one type of use.

PROBLEM 44

A 700 car parking lot is being constructed near an event site. Event parking will be subject to a fixed fee paid upon entering through a cashier lane. The cashiers are estimated to have a service rate, μ, of 270 vph for each entrance lane. The number of lanes, n, is determined by $V/(\text{PHF})\mu$, where V is the arrival rate in vph. During the hour prior to the event start, 72% of the vehicles are expected to arrive. The peak hour factor (PHF), defined similarly to the highway traffic definition of PHF, varies according to the following.

no. of lanes	PHF
1	0.70
2	0.80
3	0.87
4	0.94
5 or more	0.97

How many entrance lanes should be provided?

(A) two
(B) three
(C) four
(D) eight

Hint: A fraction of lane requires the addition of another full lane.

PROBLEM 45

A three-phase signal with 3 sec lost time per phase has the following critical movement conditions.

	phase A	phase B	phase C
volume (vph)	70	400	550
capacity (vph)	350	1350	1960

Using Webster's method, what is most nearly the recommended length of the cycle?

(A) 24 sec
(B) 26 sec
(C) 43 sec
(D) 83 sec

Hint: The optimal signal cycle, C_o, is one in which all of the phases operate so that the sums of volume/capacity ratios are slightly less than 1. By inspection, all of the phases shown have ratios less than 1.

PROBLEM 46

In the layout shown, trucks are less than 1% of vehicles using approach E and there are 20 parking maneuvers per hour. There are no buses or pedestrians. The location is near a central business district (CBD).

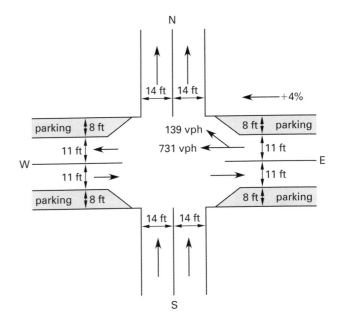

What is most nearly the saturation flow rate of approach E?

(A) 870 vph
(B) 1270 vph
(C) 1300 vph
(D) 1420 vph

Hint: The saturation flow rate modifies an ideal flow rate by approach condition adjustment factors.

PROBLEM 47

The intersection of a major two-lane, two-way street with a minor street is being evaluated by city officials for the installation of a traffic signal. The major street meets the minimum MUTCD warrant, but the minor street falls just short of the warrant. What additional action or factor could justify the installation of a traffic signal at the intersection?

(A) An engineering study indicates that a traffic signal would improve the overall safety of the intersection.
(B) 100 people use the crosswalk in the intersection when traveling to and from work at a nearby business over a 2 hr a.m. period.
(C) There have been five accidents at the intersection in the past year, each caused by the driver not stopping at the stop sign on the minor street.
(D) During the evening rush, a 15 min traffic jam occurs at the intersection when the employees of adjoining business leave work.

Hint: The motivation for installing a traffic signal needs to be supported by specific information.

PROBLEM 48

When a traffic signal is put in a flashing operation, which of the following statements is FALSE?

(A) A yellow indication should be used for the major street, and a red indication should be used for all other approaches.
(B) The most restrictive minor street should have a red indication, and all other approaches may have a yellow indication.
(C) Each approach or separate controlled turn movement shall be provided with a flashing display.
(D) Protected or protected/permissive left-turn movements shall flash red when the through movement flashes yellow.

Hint: A yellow indication does not require traffic to stop.

TRANSPORTATION PLANNING

PROBLEM 49

7000 persons commute daily to an employment center from a bedroom community, with an average commute distance of 8 mi in one direction. The average speed of the commute is 20 mph. The average car occupancy of carpools is 2.10 persons/veh. 25% of the commuters utilize carpools. The remaining arrive in single-occupant vehicles. Fuel consumption is measured by

$$F = 0.0362 \frac{\text{gal}}{\text{veh-mi}} + \frac{0.746 \frac{\text{gal}}{\text{veh-hr}}}{v}$$

There are 125,000 Btu's per gallon of fuel. Approximately how many Btu's are consumed each day for this commute pattern?

(A) 7.2×10^3
(B) 4.5×10^8
(C) 8.9×10^8
(D) 10×10^8

Hint: A daily commute includes two trips.

PROBLEM 50

A transit terminal is being developed in an urban neighborhood. One of the goals of site impact is to improve public safety. Which of the following would have the LEAST impact on this goal?

(A) improve personal security of urban travelers
(B) reduce noise and vibration impacts
(C) provide adequate lighting levels throughout walkway and platform areas
(D) improve reliability of transit service

Hint: Public safety involves minimizing anxiety about personal safety.

PROBLEM 51

The following model has been determined to show the relationship for the number of person trips per household.

$$T = 0.78 + 1.3P + 2.3A$$

T is the number of trips per household per day, P is the number of persons per household, and A is the number of autos per household.

A study zone contains 600 households, each having an average of 3.5 persons. There is an average of 2.2 autos/household. The modal split is 0.94/0.05 auto to transit, with 0.01 assigned to other. Approximately how many auto trips per day are generated by the study zone?

(A) 10 trips/day
(B) 5900 trips/day
(C) 6200 trips/day
(D) 98×10^3 trips/day

Hint: The number of auto trips is a portion of the total number of household trips.

(A) 1.35:1
(B) 2.24:1
(C) 2.48:1
(D) 2.69:1

Hint: Use the prismoidal earth volume method.

CONSTRUCTION

PROBLEM 52

On the cross sections of a roadway, the area of cut and fill has been determined for each station (see the following table). The earthwork is to be determined using the average end-area method.

station	cut (ft²)	fill (ft²)
1+00	0	50
2+00	250	70
3+00	100	0

The amount of borrow or waste between the stations is approximately

(A) 350 yd³ waste
(B) 760 yd³ borrow
(C) 760 yd³ waste
(D) 1100 yd³ borrow

Hint: The volume is a series of prisms between cross sections.

PROBLEM 53

A 3500 ft section of roadway is being excavated using the following cross section.

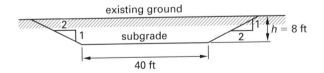

The next adjoining section requires the placement of fill; however, the mass diagram indicates that 2000 yd³ of borrow will be necessary. In order to avoid searching for borrow, the contractor proposes to lay back both side slopes of the excavation an equal amount. What will the new side slopes be?

PROBLEM 54

On a section of roadway, the cost of excavation is $2.70/yd³, the cost of overhaul is $0.90/yd³-sta, and the freehaul is 800 ft. Use the following relations.

c	cost of roadway excavation/yd³	$
h	cost of overhaul, on basis of 1 yd³/sta	$
lfh	length of freehaul	ft
loh	economical length of overhaul	ft
leh	length of economical haul	ft

What is the limit of economical haul?

(A) 33 ft
(B) 300 ft
(C) 800 ft
(D) 1100 ft

Hint: The limit of economical haul is the distance at which the contractor "breaks even" with the cost of excavation. Overhaul is the additional haul cost above the economical haul, which is not being paid by the contract.

PROBLEM 55

Hot-mix asphalt pavement for a principal arterial in an urban area will be subjected to 3% traffic growth over 15 yr. The current traffic data is as follows.

vehicle type	number of vehicles (per yr)
single units	
2-axle, 4-tire	80,000
2-axle, 6-tire	9000
3-axle or more	2000
all single units	91,000
tractor semitrailer	
5-axle	3000
all vehicles	94,000

Average Truck Factors (TF) for Different Classes of Highways and Vehicles—(United States[1])

	truck factors[2]					
	rural systems			urban systems		
vehicle type	inter-state	other principal	minor arterial	inter-state	other freeways	other principal
single-unit trucks						
2-axle, 4-tire	0.003	0.003	0.003	0.002	0.015	0.002
2-axle, 6-tire	0.21	0.25	0.28	0.17	0.13	0.24
3-axle or more	0.61	0.86	1.06	0.61	0.74	1.02
all single-units	0.06	0.08	0.08	0.05	0.06	0.09
tractor semi-trailers						
4-axle or less	0.62	0.92	0.62	0.98	0.48	0.71
5-axle[3]	1.09	1.25	1.05	1.07	1.17	0.97
6-axle or more[3]	1.23	1.54	1.04	1.05	1.19	0.90
all multiple units	1.04	1.21	0.97	1.05	0.96	0.91
all trucks	0.52	0.38	0.21	0.39	0.23	0.21

[1]Compiled from data supplied by the Highway Statistics Division, U.S. Federal Highway Administration.

[2]Individual situations may differ from these average values by 50 percent or more.

[3]Including full-trailer combinations in some states.

Reprinted with permission from *The Asphalt Handbook*, copyright © 1989, by the Asphalt Institute.

Determine the total approximate equivalent axle loading (EAL).

(A) 7300
(B) 10,500
(C) 135,000
(D) 1,750,000

Hint: Truck factors must be applied to the vehicle count in order to determine the effect of different axle loadings.

PROBLEM 56

When an asphalt concrete surface course is applied over a Type II or Type III emulsified asphalt base for light traffic conditions, what is the minimum thickness recommended by the Asphalt Institute?

(A) 1 in
(B) 2 in
(C) 2½ in
(D) 3 in

Hint: Type II and Type III base courses have slightly less strength than does a Type I base course.

PROBLEM 57

Aggregate for a roadway is to be placed in a 4 in compacted lift that is 11 ft wide. The aggregate is to be delivered in a windrow to be spread by a road grader. The windrow is to be placed 18 in high, and the top width will be one-half of the bottom width. The loose bulking factor of the aggregate is 0.68 and the in-place density is to be 135 lbm/ft³.

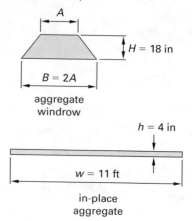

What will most nearly be the bottom width of the windrow?

(A) 2.2 ft
(B) 2.4 ft
(C) 3.3 ft
(D) 4.8 ft

Hint: The aggregate in the windrow is uncompacted bulk.

PROBLEM 58

The Superpave™ method of pavement design involves

(A) the stability of test specimens measured by the maximum load resistance (in newtons) that a specimen will develop
(B) grading of asphalt binders in which the seven-day maximum pavement design temperature and the minimum pavement design temperature are specified
(C) the ability of a test specimen to resist lateral displacement from application of a vertical load, based on density/voids analysis
(D) laboratory compaction and performance testing to achieve the maximum possible load-carrying capacity at a given design pavement temperature

Hint: The Superpave method predicts performance based on a range of expected field conditions.

PROBLEM 59

Which of the following would NOT have a marked effect on the design and construction of a hot-mix asphalt pavement surface?

(A) subsurface utility conditions
(B) alkali content of aggregate
(C) access requirements of adjoining properties
(D) availability of suitable contractors and equipment

Hint: Asphalt bonding to aggregate tolerates a very wide range of chemical compositions.

PROBLEM 60

The fineness modulus for aggregate samples is obtained by adding the cumulative percentage by weight retained on each of a specified series of sieves and then dividing the sum by 100. Determine the approximate fineness modulus of the following sample.

sieve size	percentage of sample retained, by weight (%)
3/8 in	1
no. 4	3
no. 8	11
no. 16	20
no. 30	24
no. 50	22
no. 100	17
pan	2
total	100

(A) 1.0
(B) 2.9
(C) 3.9
(D) 4.1

Hint: Percent retained must be converted to cumulative percent retained.

PROBLEM 61

Regarding the moisture content of pavement construction aggregates, which of the following statements is FALSE?

(A) Saturated surface dry (SSD) is a common design parameter used in specifying concrete mix proportions.
(B) Moisture content has no effect on the design of hot-mix asphalt mix proportions.
(C) Damp or wet aggregate contains an excess of moisture on the surface that must be considered in specifying concrete mix proportions.
(D) Air-dried aggregate has no moisture present, making it fully absorbent.

Hint: Consider the boiling point of water.

PROBLEM 62

Which of the following conditions is NOT a criterion for selecting the nominal maximum size of large aggregate to be used in a concrete pavement mix?

(A) The aggregate should not exceed a size that is the narrowest dimension of the concrete member divided by five.
(B) The aggregate should not exceed a size that is the largest necessary to keep the air-void contact below 10% of the completed mix design.
(C) The aggregate should not exceed the minimum clear spacing between reinforcing bars or mesh times 0.75.
(D) The aggregate should not exceed the depth of slab divided by three.

Hint: In order for the mix to work properly, large aggregate requires physical space within the pavement.

PROBLEM 63

Regarding aggregate properties for concrete pavements, which of the following statements is FALSE?

(A) Paste strength and aggregate bond for normal-strength concrete have a greater influence on concrete pavement strength than does aggregate strength.
(B) The abrasion resistance of aggregate is often used as a general indicator of aggregate quality.
(C) Deleterious substances are substances that enhance or strengthen the properties of aggregates.
(D) Aggregates containing higher amounts of clay or shale deteriorate more rapidly under wetting and drying conditions of pavement surfaces.

Hint: Ingredients that reduce the quality of the aggregate are undesirable.

PROBLEM 64

Financing is being prepared for rehabilitating a 10 mi long concrete highway in a rural area. The highway is 24 ft wide and is to be resurfaced with a 4 in asphalt overlay. The overlay is expected to cost $11.00 yd^2 in place. The existing concrete joints are to be repaired either by sawcutting and resealing, or by complete replacement. The joints are spaced 44 ft apart, and 15% of the joints will need to be replaced. Sawcutting and sealing will cost $450/joint, and replacement will cost $1800/joint. The life of the overlay is expected to be 7 yr. What is most nearly the annual capital recovery cost of the project, using an interest rate of 4% compounded annually?

(A) $333,000/yr
(B) $389,000/yr
(C) $402,000/yr
(D) $438,000/yr

Hint: Capital recovery assigns the initial cost to an annual amount over the expected life of the project.

PROBLEM 65

Traffic data for a section of two lane, bidirectional roadway shows the following axle load repetitions. The number of average daily trips (ADT) is 8500 with a 60%/40% directional split.

axle load grouping	gross load (lbf)	load equivalent factor	% ADT
single axle	<3000	0.0002	67
	18,000	1.00	21
2-axle	14,000	0.027	6
	22,000	0.180	6

Using thickness design criteria from *The Asphalt Handbook*, what is most nearly the equivalent axle load (EAL) for the design lane?

(A) 1900 EAL/yr
(B) 420,000 EAL/yr
(C) 440,000 EAL/yr
(D) 730,000 EAL/yr

Hint: Equivalent axle load applies a factor to each axle load group, to adjust for increased structural stress in the pavement, caused by increased axle loads.

PROBLEM 66

When applying the Marshall stability and flow tests, according to *The Asphalt Handbook*, which of the following statements is most correct.

(A) Mixes that have very low flow values and abnormally high Marshall stability values are considered too brittle to be used for flexible pavement service.
(B) Mixes with high flow values are usually more flexible and will adjust easier under traffic loads, which compensates for subgrade movement, thereby extending pavement life.
(C) Mixes with high flow values can usually be improved by reducing the asphalt content of the mix.
(D) The maximum test load applied without the sample failing is considered to be the maximum load the specimen will stand.

Hint: Marshall test procedures target a range of stability-flow values that consider durability and strength.

PROBLEM 67

According to *The Asphalt Handbook*, which of the following statements concerning asphalt pavement compaction is FALSE?

(A) Rolling should start at the low side of the spread and progress toward the high side.
(B) When rolling a thick lift, the first pass should be 12–15 in (300–375 mm) from an unsupported edge.
(C) Finish rolling may be done with steel-wheeled rollers, weighted but not vibrating.
(D) When using pneumatic-tired rollers, the tires should be kept cold to prevent asphalt from sticking to them.

Hint: Hot mat asphalt acts like a viscous, sticking mass when first laid, adhering to anything porous that cools it quickly.

GEOMETRIC DESIGN

PROBLEM 68

A specification for highway superelevation transition states that the change in cross slope can not exceed 0.02 ft/ft for each second of travel. A right-hand curve is being designed with full superelevation of 0.08 ft/ft on a four-lane divided freeway. The lanes are 12 ft wide and the design speed is 70 mph. The normal slope is 0.01 ft/ft down to the right. What is most nearly the minimum length of superelevation transition for the right side of the road?

(A) 100 ft
(B) 360 ft
(C) 410 ft
(D) 460 ft

Hint: The distance traveled in one second is the change in cross slope (superelevation) of 0.02 ft/ft.

PROBLEM 69

A specification for highway superelevation transition states that the edge profile transition is not to exceed a 1:200 difference from the centerline profile. A right-hand curve is being designed with full superelevation of 0.08 ft/ft on a four-lane divided freeway. The lanes are 12 ft wide, and the design speed is 70 mph. The normal slope is 0.01 ft/ft, and the superelevation is to be rotated about the median edge of the running lanes. What is most nearly the minimum length of transition for the right side of the roadway?

(A) 200 ft
(B) 340 ft
(C) 390 ft
(D) 2800 ft

Hint: Determine how much elevation difference will occur in the right edge of the pavement.

PROBLEM 70

Safe design speed on highway curves is determined by criteria described in the GDHS. Which of the following criteria normally apply to designing a safe curve?

    I.   curve radius
    II. passenger comfort factor
    III. sight distance
    IV. shoulder width
    V. speed-limit posting
    VI. superelevation rate
    VII. volume of traffic
    VIII. weather conditions

(A) I, II, III, and VI
(B) I, III, VI, and VIII
(C) I, II, III, VI, and VIII
(D) I, II, III, IV, V, and VIII

Hint: Safe speed on a curve balances the forces needed to keep a vehicle on the roadway against the forces tending to push the vehicle off the roadway.

PROBLEM 71

What is most nearly the north coordinate of the point of tangent (PT) shown?

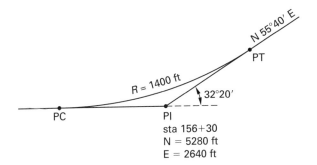

(A) 2975 ft
(B) 5500 ft
(C) 5510 ft
(D) 5615 ft

Hint: The tangent distance can be found using the radius and curve deflection. The bearing of the back tangent is not needed to solve the problem.

PROBLEM 72

For the curve shown, what is the bearing of the ahead tangent?

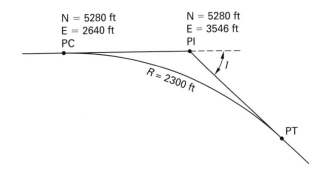

(A) 133°
(B) S 43° E
(C) S 47° E
(D) S 68°30′ E

Hint: The curve deflection can be determined using the curve radius and curve tangent length.

PROBLEM 73

A railroad curve of 6° is to be paralleled by a highway centerline, which is to be 125 ft to the inside of the railroad curve. The railroad is laid out using chord definition curves, while the highway is to use arc definition curves. What is the degree of curve for the highway?

(A) 5°18′12″
(B) 6°54′00″
(C) 6°54′13″
(D) 16°12′56″

Hint: Solve the problem using radius definition, then convert back to degree of curve.

PROBLEM 74

Superelevation transition for a 150 mph railroad is determined by the rotational change of the vehicle on U.S. standard gage track. Specifications call for an American Railway Engineering and Maintenance-of-Way Association (AREMA) design of 1.15° cross-slope change per second at the design speed. What is most nearly the transition length for a 5 in superelevation?

(A) 250 ft
(B) 660 ft
(C) 970 ft
(D) 11,000 ft

Hint: Use a standard railroad gauge of 56.5 in.

PROBLEM 75

Construction and operational characteristics of a full cloverleaf interchange between two freeways can be described by which of the following statements?

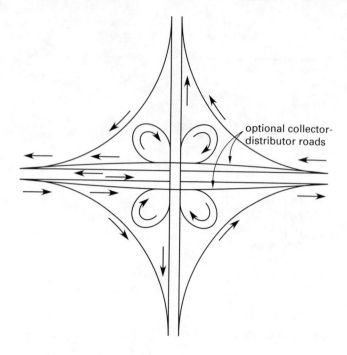

optional collector-distributor roads

(A) Only one bridge is necessary, making the interchange less costly to build.
(B) The layout is easily understood by drivers and makes the most efficient use of real estate.
(C) All movements can be made as a direct connection and at nearly the same speed as mainline travel.
(D) All ramp entrances and exits are on the right side of the roadway, but may be too closely spaced to handle higher traffic volumes in the weaving zones.

Hint: A full cloverleaf interchange has four tight circular ramps tucked between broad outer ramps and the mainline.

PROBLEM 76

An engineer is laying out a highway centerline that has a fully spiraled curve at point of intersection (PI) sta 196+00. The curve deflection is 24° Rt, and the central curve has a radius of 850 ft. Using a 250 ft spiral, what is most nearly the external distance of the curve?

(A) 19 ft
(B) 22 ft
(C) 32 ft
(D) 84 ft

Hint: The central curve of a fully spiraled curve is shifted inward to account for the offset effect of the spiral.

PROBLEM 77

A six-lane divided highway is being designed for a 60 mph running speed with a spiraled horizontal 5.73° curve. The longer of two criteria is to be used to set the length of spiral on the inside curve lanes. The first is

$$L_{s,\text{ft}} = 1.6_{(\text{ft}^2\text{-hr}^3)/\text{mi}^3} \left(\frac{v_{\text{mph}}^3}{R_{\text{ft}}} \right)$$

The other criterion is that the pavement edge profile should not deviate more than 1:200 from the centerline profile in the transition, using the spiral length as the superelevation transition. The normal slope of the roadway is 0.015 ft/ft, and the maximum superelevation is to be 0.08 ft/ft. Superelevation is to rotate about the median edge, and the lanes are 12 ft wide. What length of spiral should be used on the inside curve lanes?

(A) 350 ft
(B) 470 ft
(C) 580 ft
(D) 950 ft

Hint: Find the change in elevation of the outside edge of the pavement in relation to the centerline.

PROBLEM 78

A compound spiral transition curve is to connect between two adjacent compound horizontal circular curves. The design speed is 70 mph. The length of spiral transition is to be determined by

$$L_{s,\text{ft}} = 1.6_{(\text{ft}^2\text{-hr}^3)/\text{mi}^3} \left(\frac{v_{\text{mph}}^3}{R_{\text{ft}}} \right)$$

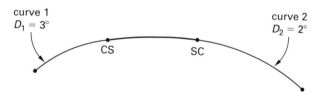

What is the approximate length of spiral required?

(A) 100 ft
(B) 190 ft
(C) 290 ft
(D) 480 ft

Hint: The change in radius from one curve to the next is related to time.

PROBLEM 79

A reverse spiral transition is to connect between two adjacent reverse horizontal circular curves on a roadway. The design speed is 60 mph. The length of spiral transition is to be determined by

$$L_{s,\text{ft}} = 1.6 \left(\frac{v^3_{\text{mi/hr}}}{R_{\text{ft}}} \right)$$

What most nearly is the total required length of spiral from the curve-to-spiral point (CS) to the spiral-to-curve point (SC)?

(A) 60 ft
(B) 120 ft
(C) 180 ft
(D) 300 ft

Hint: A spiral carries the degree of curve to zero, then from zero to the new degree of curve in the opposite direction.

PROBLEM 80

An existing simple horizontal curve is to have spirals added to the ends. The center portion of the curve is to remain unshifted. The curve has the following characteristics.

degree of curve	5°
design speed	50 mph
length of spiral transition	200 ft

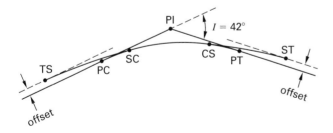

Approximately how much should the tangents be shifted in order to accommodate the spirals?

(A) −14.2 ft
(B) 1.46 ft
(C) 2.55 ft
(D) 5.82 ft

Hint: The spiral offset is the same as *p* for a spiral.

PROBLEM 81

A 3% upgrade meets a 4% downgrade on a road with a design speed of 50 mph. What is most nearly the minimum length of vertical curve required based on AASHTO recommendations for stopping sight distance?

(A) 425 ft
(B) 540 ft
(C) 590 ft
(D) 670 ft

Hint: Two situations can occur—one in which the stopping sight distance is shorter than the length of vertical curve, and the other in which the stopping sight distance is longer than the vertical curve.

PROBLEM 82

An interchange entrance ramp on a rural freeway occurs at the far end of a 1500 ft crest vertical curve and has been the scene of several rear-end accidents since it was opened. The terrain is mountainous, and the road surface is often wet in foggy conditions. At this location, the freeway has a 3.0% upgrade meeting a 2.5% downgrade. What should be the recommended posted maximum speed for the freeway mainline?

(A) 50 mph
(B) 60 mph
(C) 65 mph
(D) 70 mph

Hint: The sight distance on a crest vertical curve follows AASHTO guidelines.

PROBLEM 83

A sag vertical curve with no passing permitted is being rebuilt on a roadway in an unlit suburban area. The grade entering the curve is −6% leading into a +3.5% grade. Based on the GDHS sag vertical curve formula, what should the length of curve be, most nearly, for a speed limit of 35 mph?

(A) 130 ft
(B) 250 ft
(C) 500 ft
(D) 1000 ft

Hint: The GDHS formula is based on headlight sight distance. Only stopping sight distance should be considered, since no passing is permitted.

PROBLEM 84

Using the values given in the table, determine the maximum design speed for the following obstructed curve.

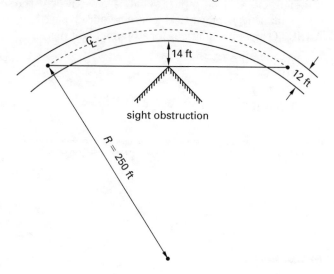

design speed (mph)	stopping-sight distance (ft)
15	75
25	150
30	200
35	250

(A) 15 mph
(B) 25 mph
(C) 30 mph
(D) 35 mph

Hint: The obstruction offset distance is the mid-ordinate of the circular curve.

TRAFFIC SAFETY

PROBLEM 85

A car skids 185 ft down a 3% grade to a stop. Friction factors are given as follows.

	v_0 (mph)			
	30	40	50	60
f	0.59	0.51	0.45	0.35

What was most nearly the speed of the car at the beginning of the skid?

(A) 39 mph
(B) 49 mph
(C) 50 mph
(D) 71 mph

Hint: Friction factors are given as averages based on the initial speed of the skid and are assumed to vary linearly.

PROBLEM 86

A car traveling at 70 mph on a 5% downgrade skids 350 ft before striking a retaining wall head-on. The coefficient of friction between the tires and the road is 0.30. What was the approximate speed of the car at impact?

(A) 35 mph
(B) 42 mph
(C) 48 mph
(D) 51 mph

Hint: A downgrade increases the required stopping distance for a given speed.

PROBLEM 87

A car traveling at 60 mph in a construction area skids into a stack of concrete barriers. Skid marks leading to the crash measure 150 ft long. The car was traveling on a 3% upgrade and experienced a pavement friction factor of 0.30. Approximately how much further would the car have skidded if it had not struck the barriers?

(A) 19 ft
(B) 210 ft
(C) 250 ft
(D) 290 ft

Hint: The friction factor is assumed to be an average over the entire skid distance.

PROBLEM 88

A car traveling on a 6% downgrade skids 60 m before colliding with another car. Police on the scene estimated the impact speed at 40 kph and determined that the pavement friction factor was 0.48. What was most nearly the speed at the beginning of the skid?

(A) 25 kph
(B) 70 kph
(C) 90 kph
(D) 95 kph

Hint: What additional distance would be required for the car to skid to a stop without a collision?

PROBLEM 89

A car brakes suddenly and skids to a stop from 60 mph. The car initially skids 150 ft on pavement with a friction factor of 0.30. The skid continues onto wet grass on hard soil with a friction factor of 0.10. Both parts of the skid are on a 3% upgrade. Approximately how long is the skid on the grassy surface?

(A) 540 ft
(B) 750 ft
(C) 1300 ft
(D) 1600 ft

Hint: An upgrade decreases the length of skid from a given speed.

PROBLEM 90

Accident data has been tabulated for a 20 mi section of an arterial highway.

accident type	1996	1997	1998	1999
fatal	0	3	2	5
personal injury	35	35	50	50
property damage only	130	180	190	230
ADT (veh)	15,500	16,000	16,500	17,000

What is most nearly the accident rate for injury accidents per 100 million vehicle miles (HMVM)?

(A) 0.038 accident/HMVM
(B) 36 accidents/HMVM
(C) 38 accidents/HMVM
(D) 190 accidents/HMVM

Hint: Injury accidents include personal injury and fatal accidents.

PROBLEM 91

A major highway segment through a business district 2.5 mi long is being evaluated for accident rates over a 4 yr period. The average daily traffic (ADT) is 28,000 veh/day. During the study period there have been 9 accidents/yr involving death or injury. The statewide statistic for similar types of roadways is 150 accidents per 100 million vehicle miles (HMVM) involving death or injury. The business district evaluation uses a rate quality control method in which the critical accident rate for a segment is determined by

$$R_{\text{crit}} = R_{\text{ave}} + K\sqrt{\frac{R_{\text{ave}}}{\text{traffic base}}}$$

A confidence level of 95% is being assigned to the study.

confidence level (%)	K
90	1.282
95	1.645
98	2.054

The relationship appoximates one standard error, and the units do not necessarily balance. However, the accident rate and the traffic base use the same magnitude of vehicle miles. What is most nearly the accident rate of this segment compared to the statewide critical rate?

(A) 0.14
(B) 0.19
(C) 0.88
(D) 5.4

Hint: The confidence level is selected for the probability that the accident rate is more than a random occurrence.

PROBLEM 92

A crash cushion is being designed for a 70 mph impact. The design criterion calls for a maximum deceleration rate of $7g$ with 75% efficiency. What is the minimum compression length of the crash cushion?

(A) 15 ft
(B) 17 ft
(C) 23 ft
(D) 31 ft

Hint: The efficiency is comparable to a design factor of safety.

PROBLEM 93

An energy absorption barrier is being placed against a backwall. The barrier is to absorb the impact of a 4500 lbm vehicle approaching at a speed of 70 mph. Use a factor of safety of 1.5 and a deceleration rate no greater than $7.5g$. The vehicle is to stop in 23.4 ft.

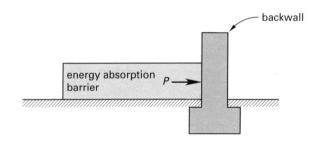

What is most nearly the force on the backwall?

(A) 32 kips
(B) 47 kips
(C) 51 kips
(D) 1500 kips

Hint: The impact force is the deceleration force.

PROBLEM 94

A highway intersection with a crossroad has been the site of many left-turn and head-on collisions recently. What is the LEAST likely cause of the accidents?

(A) the large volume of left turns
(B) inadequate roadway lighting
(C) the absence of a special left-turning phase
(D) inadequate gaps in traffic

Hint: Left turns involve more critical space allocation than does any other intersection move.

PROBLEM 95

A signalized intersection has had a large number of accidents due to slipery surface conditions. Which of the following countermeasures would have the greatest likelihood of reducing accidents?

(A) reduce the speed limit on approaches
(B) install advance overhead signals
(C) improve roadway lighting
(D) prohibit turns

Hint: Poor surface conditions result in reduced traction for turning and stopping.

PROBLEM 96

The Federal Highway Administration (FHWA) and National Highway Traffic Safety Administration (NHTSA) have made efforts to encourage states to increase emphasis on and find support for safety countermeasures. Which of the following actions would NOT be part of such a program?

(A) Provide improved signing, marking, and delineation.
(B) Increase moving violation enforcement through work zones.
(C) Improve ergonomic design of vehicle interiors.
(D) Establish pedestrian safety as a priority area.

Hint: FHWA and NHTSA deal primarily with on-highway activities.

PROBLEM 97

A traffic signal system is to be installed at the entrance to a shopping center. The signal system costs $120,000 to install and is expected to have a salvage value of $15,000 at the end of its 15 yr life. The initial cost is to be depreciated using the straight-line method. If interest will be 6%, what is most nearly the value of the system at the end of the tenth year?

(A) $40,000
(B) $51,000
(C) $61,000
(D) $70,000

Hint: The remaining value includes the undepreciated initial cost, plus the tenth-year salvage value.

PROBLEM 98

An escrow fund has been established in the amount of $2.5 million for demolition of a railroad crossing in 15 yr. The fund is expected to yield 4.5% interest compounded annually. The value of the fund at the end of the term will be most nearly

(A) $2,340,000
(B) $4,190,000
(C) $4,690,000
(D) $4,840,000

Hint: The nominal interest is the effective annual interest.

PROBLEM 99

A transit terminal in a small town is to be built using $500,000 of a bond issue. The bonds will yield 6% annually and will be retired at the end of 10 yr. A sinking fund is to be established to retire the bonds and pay the interest, into which 10 annual end-of-year payments will be made. The sinking fund will return a rate of 4.5% annually. The cost of selling the bonds is 2% of the issue. What is most nearly the annual amount necessary for payment into the sinking fund?

(A) $31,000
(B) $42,000
(C) $69,000
(D) $72,000

Hint: The cost of the bond issue is added to the principal amount.

PROBLEM 100

The following list of roadway segment deficiencies has been prepared for a Transportation Systems Management (TSM) study.

I. lane width under 11 ft
II. signs and pavement markings not in conformance with ITE/AASHTO standards
III. peak average travel speed less than 20 mph
IV. roadway drainage does not meet FHWA standards
V. sight distance not in conformance with MUTCD
VI. roadway level of service lower than C
VII. roadway lighting levels below Illuminating Engineering Society (IES) standards
VIII. high roadway accident rates
IX. too many delivery truck companies servicing business establishments

Which of the deficiencies are LEAST likely to qualify for TSM improvement programs?

(A) III, VII, and IX
(B) IV, VII, and IX
(C) V, VI, and VIII
(D) V, VII, and IX

Hint: TSM programs create efficient use of road space through traffic management programs.

Breadth Solutions

TRAFFIC ANALYSIS

SOLUTION 1

To obtain the average distance from car to car, apply an instant of time to calculate the number of cars that will be in a one-lane mile of roadway during that instant.

Solve for density.

$$D = \frac{V_p}{v} = \frac{2200 \frac{\text{pc}}{\text{hr-ln}}}{42 \frac{\text{mi}}{\text{hr}}}$$

$$= 52.4 \text{ pcpmpl}$$

Find the spacing between cars in one lane.

$$\text{spacing} = \frac{1 \text{ mi-lane}}{D} = \frac{5280 \frac{\text{ft}}{\text{mi-lane}}}{52.4 \frac{\text{pc}}{\text{mi-lane}}} = 100.8 \text{ ft/pc}$$

$$\text{distance between cars} = \text{spacing} - \text{average car length}$$
$$= 100.8 \text{ ft} - 19 \text{ ft}$$
$$= 81.8 \text{ ft} \quad (82 \text{ ft})$$

The answer is (B).

Alternate Solution:

$$\text{spacing} = \frac{1 \text{ mi-lane} - DL}{D}$$

$$= \frac{(1 \text{ mi-lane})\left(5280 \frac{\text{ft}}{\text{mi-lane}}\right) - \left(52.4 \frac{\text{pc}}{\text{mi-lane}}\right)\left(19 \frac{\text{ft}}{\text{pc}}\right)}{52.4 \frac{\text{pc}}{\text{mi-lane}}}$$

$$= 81.8 \text{ ft} \quad (82 \text{ ft})$$

Why Other Options Are Wrong

(A) This incorrect solution assumes it is necessary to convert speed to feet per second, then directly subtracts the car length. This gives the length of travel not occupied by a car in one second instead of the actual distance between cars. It is necessary to carry the solution further to determine the following distance using a density relationship.

(C) This solution is incorrect because 148 ft is the spacing from the front of one car to the front of the next car. The length of the car must be subtracted from the spacing to find the distance between cars.

(D) This solution is incorrect because an extra conversion from miles per hour to feet per second was inserted in the first equation.

SOLUTION 2

The total volume, which consists of a mix of vehicle types, must be converted to equivalent passenger-car volume by assigning passenger-car equivalents to the trucks, buses, and RVs. The heavy vehicle factor is determined by HCM Eq. 23-3 (see HCM Exh. 23-8 for passenger-car equivalents).

$$f_{HV} = \frac{1}{1 + P_T(E_T - 1) + P_R(E_R - 1)}$$

Buses and trucks are counted together as heavy vehicles when the percentage of heavy vehicles is at least five times the percentage of RVs. The RV count is added to the heavy vehicle count so that all heavy vehicles are counted as trucks.

$$f_{HV} = \frac{1}{1 + P_{HV}(E_{HV} - 1)} = \frac{1}{1 + (0.07)(2.5 - 1)}$$
$$= 0.905$$

The equivalent passenger-car flow rate is determined by HCM Eq. 23-2. The driver population factor is 1.0 for predominantly commuter traffic.

$$V_p = \frac{V}{(\text{PHF})(\text{no. of lanes})f_{HV}f_p}$$

$$= \frac{1970 \frac{\text{veh}}{\text{hr}}}{(0.85)(4 \text{ lanes})(0.905)(1.0)}$$

$$= 640 \text{ pcphpl} \quad (660 \text{ pcphpl})$$

The answer is (D).

Why Other Options Are Wrong

(A) This incorrect solution would result from misplacing the peak hour factor (PHF) in the numerator.

(B) This erroneous solution would occur if the correction for passenger-car equivalents were not included in the volume adjustment.

(C) This erroneous solution would occur if the passenger-car equivalents were selected for level terrain instead of for rolling terrain.

SOLUTION 3

The heavy vehicle factor is determined by HCM Eq. 23-3. Trucks and buses are combined because of similar performance and size.

$$f_{HV} = \frac{1}{1 + P_T(E_T - 1) + P_R(E_R - 1)}$$
$$= \frac{1}{1 + (0.03 + 0.02)(4.5 - 1) + (0.03)(4.0 - 1)}$$
$$= 0.791$$

The equivalent passenger flow rate is found using HCM Eq. 23-2. The driver population factor is given as 0.95, which is appropriate for traffic comprised of mostly commuters. The PHF is given as 0.90.

$$V_p = \frac{V}{(\text{PHF})(\text{no. of lanes}) f_{HV} f_p}$$
$$= \frac{4495 \frac{\text{veh}}{\text{hr}}}{(0.90)(3 \text{ lanes})(0.791)(0.95)}$$
$$= 2215 \text{ pcphpl}$$

Since no other information is given, use the design speed as the base free-flow speed (BFFS), and determine the adjusted free-flow speed using HCM Eq. 23-1.

$$\text{FFS} = \text{BFFS} - f_{LW} - f_{LC} - f_N - f_{ID}$$
$$= 60 \text{ mph} - 1.9 - 0.0 - 3.0 - 0.0$$
$$= 55.1 \text{ mph}$$

Level of service (LOS) is determined by the density in passenger cars per mile per lane.

$$D = \frac{V_p}{\text{speed}} = \frac{2215 \frac{\text{pc}}{\text{hr-lane}}}{55.1 \frac{\text{mi}}{\text{hr}}}$$
$$= 40.2 \text{ pcpmpl}$$

From HCM Exh. 23-2, based upon the density range of 35–45 pcpmpl, the LOS is E.

The answer is (C).

Why Other Options Are Wrong

(A) This incorrect solution could result from improper use of the peak hour factor (PHF). According to HCM Exh. 23-2, in order to operate at LOS C, there would have to be less than 26 pcpmpl.

(B) In order to operate at LOS D, the maximum density would have to have been found to be less than 35 pcpmpl. This density can be incorrectly found by not including the heavy vehicle and driver population factors.

(D) The maximum density for LOS E is 45 pcpmpl. Should the setting be incorrectly interpreted as rural because of the nearby recreatonal park and the PHF selected as 0.80, the density would be determined to be LOS F.

SOLUTION 4

A criteria of level of service (LOS) is the maximum service flow rate per hour per lane of highway. The service flow rate is the passenger-car equivalent flow at a free-flow speed. The flow rate must be determined in passenger-car equivalents of the total traffic vehicle mix using HCM Eq. 21-3.

$$V_p = \frac{V}{(\text{PHF})(\text{no. of lanes}) f_{HV} f_p}$$

Determine the heavy vehicle factor using HCM Eq. 21-4. HCM Exh. 21-8 or 23-8 shows car equivalents of heavy vehicles for extended freeway segments. There are no RVs in this case; therefore, only the truck part of the equation is considered.

$$f_{HV} = \frac{1}{1 + P_T(E_T - 1)}$$
$$= \frac{1}{1 + (0.07)(1.5 - 1)}$$
$$= 0.966$$

Determine the service flow rate. The driver population factor is taken as 1.00 if no other information is available.

$$V_p = \frac{2540 \frac{\text{veh}}{\text{hr}}}{(0.92)(2 \text{ lanes})(0.966)(1)}$$
$$= 1429 \text{ pcphpl}$$

Determine the adjusted free-flow speed using HCM Eq. 21-1, based on the restrictions shown.

$$\text{FFS} = \text{BFFS} - F_{LW} - F_{LC} - F_M - F_A$$

The design speed is 60 mph; therefore, the base free-flow speed (BFFS) is 60 mph.

The median is divided; therefore (from HCM Exh. 21-6), $F_M = 0$ mph.

The lane width is 10 ft; therefore (from HCM Exh. 21-4), $F_{LW} = 6.6$ mph.

The lateral clearance is 6 ft total; therefore (from HCM Exh. 21-5), $F_{LC} = 1.3$ mph.

There are access points every 300 ft (which equates to between 10 and 20 per mile); therefore (from HCM Exh. 21-7), $F_A = 5.0$ mph.

$$\text{FFS} = 60\,\frac{\text{mi}}{\text{hr}} - 0\,\frac{\text{mi}}{\text{hr}} - 6.6\,\frac{\text{mi}}{\text{hr}}$$
$$\quad - 1.3\,\frac{\text{mi}}{\text{hr}} - 5.0\,\frac{\text{mi}}{\text{hr}}$$
$$= 47.1 \text{ mph}$$

For a free-flow speed of 47 mph and a service flow rate of 1429 pcphpl, the LOS is D using HCM Exh. 21-2.

The answer is (B).

Why Other Options Are Wrong

(A) This answer is incorrect because a service flow rate of 1429 pcphpl would be LOS C at a free-flow speed (FFS) of 60 mph. This answer could only be selected if the FFS were not reduced by the clearance factors.

(C) This answer is incorrect because selection of a PHF that is too low will result in too high of a design hour volume. Using an average running speed of 47 mph, this volume would appear to be LOS E.

(D) This answer is incorrect because not dividing the adjusted flow volume by two lanes would result in a service flow rate that would appear to be LOS F, or jam density.

SOLUTION 5

The hourly vehicle volume is a count of the general mix of vehicle types, usually reported in units of vehicles per hour. The design hourly flow rate must reflect the influence of heavy vehicles, the hourly variation of traffic, and the characteristics of the driver population. The design equivalent passenger-car flow rate is calculated using the heavy-vehicle and peak-hour adjustment factors and includes the driver population adjustment.

The answer is (C).

Why Other Options Are Wrong

(A) This incorrect option does not convert to passenger-car equivalents.

(B) This incorrect option is an inaccurate definition of peak hour factor.

(D) This incorrect option does not correct for passenger-car equivalents.

CONSTRUCTION

SOLUTION 6

The basic equation is

$$\text{required fill} = \text{available on-site material} + \text{borrow material}$$

Assume all of the 24,000 yd³ of on-site earth will be used.

Determine how much fill the on-site material will make. In the volume adjustment factor, s is positive for swell and negative for shrinkage.

$$V_{F,\text{on-site}} = V_{C,\text{on-site}}(1 + s)$$
$$= (24{,}000 \text{ yd}^3)(1 + (-0.10))$$
$$= 21{,}600 \text{ yd}^3$$

The remainder of the fill must be obtained from the borrow pit.

$$V_{F,\text{borrow}} = V_{F,\text{total}} - V_{F,\text{on-site}}$$
$$= 30{,}000 \text{ yd}^3 - 21{,}600 \text{ yd}^3$$
$$= 8400 \text{ yd}^3$$

Find the volume of borrow cut required.

$$V_{C,\text{borrow}} = \frac{V_{F,\text{borrow}}}{1 + (-0.10)} = \frac{8400 \text{ yd}^3}{1 - 0.10}$$
$$= 9333 \text{ yd}^3$$

Find the unit mass of hauled material, using the bulking factor to compensate for the volume increase.

$$\text{unit mass} = \left(\frac{M}{V}\right)\left(\frac{1}{1+s}\right)$$
$$= \left(120\,\frac{\text{lbm}}{\text{ft}^3}\right)\left(\frac{1}{1+0.18}\right)$$
$$= 101.7 \text{ lbm/ft}^3$$

Check the capacity of the truck.

$$\frac{48{,}000 \text{ lbm}}{15 \text{ yd}^3} = \left(3200\,\frac{\text{lbm}}{\text{yd}^3}\right)\left(\frac{1 \text{ yd}^3}{27 \text{ ft}^3}\right)$$
$$= 118.5 \text{ lbm/ft}^3$$

Since the truck capacity of 118.5 lbm/ft³ is greater than the unit mass of the hauled material, 101.7 lbm/ft³, the maximum volume of the truck will limit the truck's load. Determine the hauling volume of the borrow material.

$$V_{\text{haul}} = V_{\text{borrow}}(1+s)$$
$$= (9333 \text{ yd}^3)(1+0.18)$$
$$= 11{,}013 \text{ yd}^3$$

$$\text{no. of loads} = \frac{V_{\text{haul}}}{\text{truck capacity}} = \frac{11{,}013 \text{ yd}^3}{15 \text{ yd}^3}$$
$$= 734.2 \quad (734)$$

The answer is (D).

Why Other Options Are Wrong

(A) Dividing instead of multiplying the on-site material by the shrinkage factor would overestimate the amount of borrow needed, as in this incorrect answer.

(B) Adjusting for the bulking factor for hauling but ignoring shrinkage results in this incorrect answer.

(C) This incorrect answer assumes the truck can carry the full 48,000 lbf instead of being limited by volume. If this were true, the truck would be overfilled and there would be too few truckloads.

SOLUTION 7

The required quantity of material is based on in-place compacted volume. The cost of hauling is based on weight. The bulking factor determines the bulk weights of the unit volumes of material to be transported from each source.

Find the bulk weight per cubic yard of sand, for hauling.

$$\text{bulk weight} = \left(\begin{array}{c} \text{source in-} \\ \text{place spec. wt.} \end{array} \right) \left(\frac{1}{1 + \text{bulking factor}} \right)$$

For source I,

$$\text{bulk weight} = \left(130 \frac{\text{lbf}}{\text{ft}^3} \right)\left(27 \frac{\text{ft}^3}{\text{yd}^3} \right)\left(\frac{1}{1+0.11} \right)$$
$$= \left(3162 \frac{\text{lbf}}{\text{yd}^3} \right)\left(\frac{1 \text{ ton}}{2000 \text{ lbf}} \right)$$
$$= 1.581 \text{ ton/yd}^3$$

For source II,

$$\text{bulk weight} = \left(130 \frac{\text{lbf}}{\text{ft}^3} \right)\left(27 \frac{\text{ft}^3}{\text{yd}^3} \right)\left(\frac{1}{1+0.17} \right)$$
$$= \left(3000 \frac{\text{lbf}}{\text{yd}^3} \right)\left(\frac{1 \text{ ton}}{2000 \text{ lbf}} \right)$$
$$= 1.500 \text{ ton/yd}^3$$

For source III,

$$\text{bulk weight} = \left(130 \frac{\text{lbf}}{\text{ft}^3} \right)\left(27 \frac{\text{ft}^3}{\text{yd}^3} \right)\left(\frac{1}{1+0.12} \right)$$
$$= \left(3134 \frac{\text{lbf}}{\text{yd}^3} \right)\left(\frac{1 \text{ ton}}{2000 \text{ lbf}} \right)$$
$$= 1.567 \text{ ton/yd}^3$$

Find the cost per cubic yard of sand delivered to the site.

$$\text{cost} = \left(\left(\begin{array}{c} \text{bulk} \\ \text{weight} \end{array} \right)\left(\begin{array}{c} \text{cost} \\ \text{per ton} \end{array} \right) \right)$$
$$+ \left(\left(\begin{array}{c} \text{bulk} \\ \text{weight} \end{array} \right)\left(\begin{array}{c} \text{hauling cost} \\ \text{per ton} \end{array} \right)\left(\begin{array}{c} \text{haul} \\ \text{distance} \end{array} \right) \right)$$

For source I,

$$\text{cost} = \left(\left(1.581 \frac{\text{ton}}{\text{yd}^3} \right)\left(\frac{\$5.40}{1 \text{ ton}} \right) \right)$$
$$+ \left(\left(1.581 \frac{\text{ton}}{\text{yd}^3} \right)\left(\frac{\$0.23}{1 \text{ ton-mi}} \right)(20 \text{ mi}) \right)$$
$$= \$15.81/\text{yd}^3$$

For source II,

$$\text{cost} = \left(\left(1.500 \frac{\text{ton}}{\text{yd}^3} \right)\left(\frac{\$5.70}{1 \text{ ton}} \right) \right)$$
$$+ \left(\left(1.500 \frac{\text{ton}}{\text{yd}^3} \right)\left(\frac{\$0.23}{1 \text{ ton-mi}} \right)(17 \text{ mi}) \right)$$
$$= \$14.42/\text{yd}^3$$

For source III,

$$\text{cost} = \left(\left(1.567 \frac{\text{ton}}{\text{yd}^3} \right)\left(\frac{\$8.10}{\text{ton}} \right) \right)$$
$$+ \left(\left(1.567 \frac{\text{ton}}{\text{yd}^3} \right)\left(\frac{\$0.23}{1 \text{ ton-mi}} \right)(12 \text{ mi}) \right)$$
$$= \$17.02/\text{yd}^3$$

Source II is the lowest cost.

The answer is (B).

Why Other Options Are Wrong

(A) In this incorrect answer, source I appears to cost less due to improper multiplication by the bulking factor (instead of division).

For source I,

$$\text{bulk weight} = \left(\begin{array}{c}\text{source in-}\\\text{place spec. wt.}\end{array}\right)\left(\frac{1}{1+\text{bulking factor}}\right)$$

$$= \left(130\,\frac{\text{lbf}}{\text{ft}^3}\right)\left(27\,\frac{\text{ft}^3}{\text{yd}^3}\right)\left(\frac{1}{1+0.11}\right)$$

$$= \left(3896\,\frac{\text{lbf}}{\text{yd}^3}\right)\left(\frac{1\text{ ton}}{2000\text{ lbf}}\right)$$

$$= 1.948\text{ ton/yd}^3$$

$$\text{cost} = \left(\left(\begin{array}{c}\text{bulk}\\\text{weight}\end{array}\right)\left(\begin{array}{c}\text{cost}\\\text{per ton}\end{array}\right)\right)$$
$$+ \left(\left(\begin{array}{c}\text{bulk}\\\text{weight}\end{array}\right)\left(\begin{array}{c}\text{hauling cost}\\\text{per ton}\end{array}\right)\right.$$
$$\left.\times \left(\begin{array}{c}\text{haul}\\\text{distance}\end{array}\right)\right)$$

$$= \left(\left(1.948\,\frac{\text{ton}}{\text{yd}^3}\right)\left(\frac{\$5.40}{1\text{ ton}}\right)\right)$$
$$+ \left(\left(1.948\,\frac{\text{ton}}{\text{yd}^3}\right)\left(\frac{\$0.23}{\text{ton-mi}}\right)(20\text{ mi})\right)$$

$$= \$19.48/\text{yd}^3$$

For source II,

$$\text{bulk weight} = \left(130\,\frac{\text{lbf}}{\text{ft}^3}\right)\left(27\,\frac{\text{ft}^3}{\text{yd}^3}\right)\left(\frac{1}{1+0.17}\right)$$

$$= \left(4107\,\frac{\text{lbf}}{\text{yd}^3}\right)\left(\frac{1\text{ ton}}{2000\text{ lbf}}\right)$$

$$= 2.054\text{ ton/yd}^3$$

$$\text{cost} = \left(\left(2.054\,\frac{\text{ton}}{\text{yd}^3}\right)\left(\frac{\$5.70}{1\text{ ton}}\right)\right)$$
$$+ \left(\left(2.054\,\frac{\text{ton}}{\text{yd}^3}\right)\left(\frac{\$0.23}{\text{ton-mi}}\right)(17\text{ mi})\right)$$

$$= \$19.74/\text{ton}$$

(C) This answer is incorrect. Source III is significantly higher in cost than either of the other sources, in spite of its closer proximity to the job site.

(D) This incorrect answer ignores the cost of hauling, making it seem that the cost per cubic yard is most nearly the same for source I and source II.

For source I,

$$\text{cost} = \left(\left(\begin{array}{c}\text{bulk}\\\text{weight}\end{array}\right)\left(\begin{array}{c}\text{cost}\\\text{per ton}\end{array}\right)\right)$$
$$+ \left(\left(\begin{array}{c}\text{bulk}\\\text{weight}\end{array}\right)\left(\begin{array}{c}\text{hauling cost}\\\text{per ton}\end{array}\right)\left(\begin{array}{c}\text{haul}\\\text{distance}\end{array}\right)\right)$$

$$= \left(1.581\,\frac{\text{ton}}{\text{yd}^3}\right)\left(\frac{\$5.40}{\text{ton}}\right)$$

$$= \$8.54/\text{yd}^3$$

For source II,

$$\text{cost} = \left(1.500\,\frac{\text{ton}}{\text{yd}^3}\right)\left(\frac{\$5.70}{\text{ton}}\right)$$

$$= \$8.55/\text{yd}^3$$

SOLUTION 8

Testing for air entrainment is used for concrete materials. Air entrainment is employed to improve frost damage resistance in concrete. Therefore, test IV is not applicable to asphalt. Water is used in portland cement mixes for hydration, but water would not be present in the higher temperatures of asphalt mixes. Therefore, test VII is not applicable. Tests for surface tension are not relevant to asphalt materials, in that water is not used in hot-mix asphalt pavements. (Surface tension and miscibility are important in portland cement mixes, in order to control hydration.) Therefore, test IX is not applicable to asphalt.

The tests not applicable to asphalt are IV, VII, and IX.

The answer is (C).

Why Other Options Are Wrong

(A) This answer is incorrect. A flash point test (Test VIII) is important for asphalt, to determine the safety of the mix at working temperatures.

(B) This answer is incorrect. Viscosity and distillation tests (Tests III and X) are typical for asphaltic cement materials. Viscosity is a measure of the fluidity, or flow properties, of the asphalt cement at various working temperatures. Distillation of a sample removes volatile elements at controlled temperatures to determine the quantity and quality of asphalt ingredients.

(D) This answer is incorrect. Ductility (Test V) is a test performed on asphalt to determine its ability to withstand traffic without fracturing or breaking up. A flash-point test (Test VIII) is important, for asphalt materials, in order to determine the safety of handling the heated material during mixing processes.

SOLUTION 9

Where there are multiple paths between subsets of activities, analyze the subsets to find the longest time path, then string together the longest subset paths to complete the critical path from start to finish. To clarify the situation, sketch the project network, showing activities, durations, and the critical path.

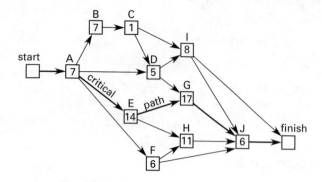

The critical path is as follows: start, A, E, G, J, finish.

$$\text{time required} = 7 \text{ days} + 14 \text{ days}$$
$$+ 17 \text{ days} + 6 \text{ days}$$
$$= 44 \text{ days}$$

The answer is (C).

Why Other Options Are Wrong

(A) This answer is incorrect. Path start, A, D, I, J, finish (26 days) is quicker than the critical path, but leaves out the full time required for both F and G activities.

(B) This answer is incorrect. Path start, A, B, C, I, J, finish (29 days) is longer than the quickest path, but still is not enough time for both F and G activities.

(D) This answer is incorrect, since there is no path longer than 44 days.

SOLUTION 10

The difference between EF and LS between *each event* is called "float." The longest duration of *parallel* activities, or all activities for which EF and LS coincide, determines the critical path. There is no slack time, or "float," along the critical path. Because there is no float along the critical path, ES and LF are determined by the activity durations.

The answer is (D).

Why Other Options Are Wrong

(A) This answer is incorrect because there is no "float" on the critical path.

(B) This answer is incorrect. Long-duration activities may be paralleled by a series of activities that are shorter in duration; but when performed in sequence, these short-duration activities total more time than the long-duration activities.

(C) This answer is incorrect because there is no slack time on the critical path.

GEOMETRIC DESIGN

SOLUTION 11

Find the PC station of the unshifted curve. First, determine the curve tangent length.

$$T = R_1 \tan \frac{I}{2}$$
$$= (1430 \text{ ft})(0.127216)$$
$$= 181.92 \text{ ft}$$

Subtract the curve tangent from the PI station to find the PC station.

$$\begin{array}{r} \text{PI sta } 572{+}00 \\ -181.92 \\ \hline \text{PC sta } 570{+}18.08 \end{array}$$

Find the length of the new tangent. Let δ be the added length to extend to the shifted position of the ahead tangent.

$$\delta T = \frac{\text{offset}}{\sin I} = \frac{12.00 \text{ ft}}{0.250380}$$
$$= 47.93 \text{ ft}$$
$$\text{new } T = T + \delta T = 181.92 \text{ ft} + 47.93 \text{ ft}$$
$$= 229.85 \text{ ft}$$
$$R_2 = \frac{\text{new } T}{\tan \frac{I}{2}} = \frac{229.85 \text{ ft}}{0.127216}$$
$$= 1806.77 \text{ ft} \quad (1810 \text{ ft})$$

The answer is (C).

Why Other Options Are Wrong

(A) This incorrect answer is the radius if the ahead tangent is shifted to the right.

(B) This incorrect answer is the radius if cosine (instead of sine) is used to find the tangent extension.

(D) This incorrect answer is the radius if the deflection angle is entered into the calculation as 14.3° instead of 14.5°.

SOLUTION 12

The sum of the two curve arc deflections α and β equals the total curve arc deflection.

Let C, D, and E be the points of tangency.

$$CA = AD = R \tan \frac{\alpha}{2}$$
$$DB = BE = R \tan \frac{\beta}{2}$$

Combining,

$$AB = AD + DB$$
$$= R\left(\tan\frac{\alpha}{2} + \tan\frac{\beta}{2}\right)$$
$$= 330 \text{ ft}$$

Rearranging,

$$R = \frac{AB}{\tan\dfrac{\alpha}{2} + \tan\dfrac{\beta}{2}}$$
$$= \frac{330 \text{ ft}}{0.221695 + 0.344328}$$
$$= 583.02 \text{ ft} \quad (580 \text{ ft})$$

The answer is (C).

Why Other Options Are Wrong

(A) This incorrect answer would be the radius if the angles were not divided by two when calculating the radius.

(B) This answer incorrectly assumes that the long-chord formula would apply, using line AB as the chord.

(D) By inadvertently multiplying the tangent values in the denominator instead of adding them, the radius formula produces an overly large radius.

SOLUTION 13

Construct a triangle by connecting the center, point O, to point P. Then drop an altitude from point P, intersecting base AO at point E.

$$AG = x = EP$$
$$GP = y$$
$$OE = OA - y$$

From right triangle OEP,

$$OE = \sqrt{R^2 - x^2}$$

Find the curve radius.

$$D = \frac{(180°)(100)}{\pi R}$$
$$R = \frac{5729.58}{D} = \frac{5729.58°\text{-ft}}{2°}$$
$$= 2864.79 \text{ ft}$$

$$OE = \sqrt{(2864.79 \text{ ft})^2 - (200 \text{ ft})^2}$$
$$= 2857.80 \text{ ft}$$
$$y = OA - OE$$
$$= 2864.79 \text{ ft} - 2857.80 \text{ ft}$$
$$= 6.99 \text{ ft} \quad (7.0 \text{ ft})$$

The answer is (C).

Why Other Options Are Wrong

(A) Determining the curve radius by incorrectly applying the conversion from degree of curve to radius will result in a radius that is much too large.

(B) This incorrect answer results from confusing the curve deflection to point P with the degree of curve. The deflection from AG to the chord AP is one half of the angle AOP, incorrectly assumed to be 2°.

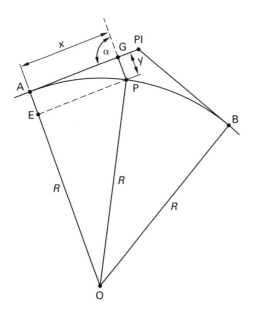
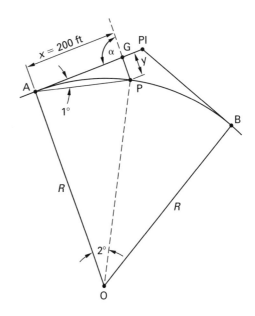

$$\text{GAP} = \frac{\text{AOP}}{2} = 1°$$

$$\tan(\text{GAP}) = \frac{y}{x}$$

$$y = x \tan 1°$$
$$= (200 \text{ ft}) \tan 1°$$
$$= 3.49 \text{ ft}$$

(D) This incorrect answer is the result of using the long chord and mid-ordinate formulas to solve for y and incorrectly doubling the deflection angle.

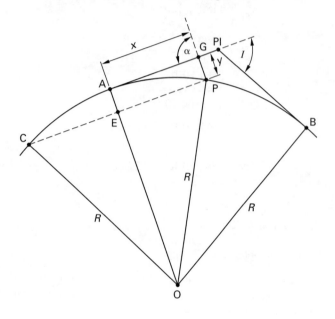

Extend the curve AP and line PE back to meet at point C. Line CP is the long chord of curve CP.

$$\text{CP} = 2(\text{AG}) = (2)(200 \text{ ft})$$
$$= 400 \text{ ft}$$
$$= 2R \sin \frac{I}{2}$$

$$\sin \frac{I}{2} = \frac{\text{CP}}{2R} = \frac{400 \text{ ft}}{(2)(2864.79 \text{ ft})}$$
$$= 0.0698$$

$$\frac{I}{2} = 4.003°$$
$$I = 8.006°$$

Here is where it is easy to mistake 8.006° for AOP and mistakenly assume that it must be doubled for the next part to represent COP.

$$M = R(1 - \cos 8.006°)$$
$$= (2864.79)(1 - \cos 8.006°)$$
$$= 27.9 \text{ ft}$$

SOLUTION 14

With the deflection and external known, solve for the first radius. Use the formula that relates the external to the curve radius and curve deflection.

$$E = R \operatorname{exsec} \frac{I}{2}$$

Rearranging,

$$R = \frac{E}{\dfrac{1}{\cos \dfrac{I}{2}} - 1} = \frac{35 \text{ ft}}{\dfrac{1}{\cos 33.875°} - 1}$$
$$= 171.192 \text{ ft}$$

Find the degree of curve.

$$D = \frac{(180°)(100)}{\pi R} = \frac{(180°)(100 \text{ ft})}{\pi(171.192 \text{ ft})}$$
$$= 33.469°$$

The nearest whole degree is 33°; therefore, find the radius for a 33° curve.

$$R = \frac{5729.58}{D} = \frac{5729.578°\text{-ft}}{33°}$$
$$= 173.624 \text{ ft} \quad (170 \text{ ft})$$

The answer is (B).

Why Other Options Are Wrong

(A) Incorrectly determining the secant from the inverse of the sine instead of from the inverse of the cosine will result in this incorrect answer.

(C) This incorrect answer is the radius before rounding the deflection to an even 33°.

(D) Incorrectly using the formula for the mid-ordinate instead of the external results in this answer.

SOLUTION 15

Solve for the length of the curve tangent.

$$T = R \tan \frac{I}{2} = (2500 \text{ ft}) \tan \frac{35°}{2}$$
$$= (2500 \text{ ft})(0.315299)$$
$$= 788.25 \text{ ft}$$

Find the station of the point of curvature (PC).

```
PI sta 485+26.75
  -       7 88.25
PC sta 477+38.5
```

The length of the curve is added to the PC station to determine the station of the point of tangent (PT). Solve for length of curve.

$$L = RI\left(\frac{2\pi}{360°}\right) = (2500 \text{ ft})(35°)\left(\frac{2\pi}{360°}\right)$$
$$= 1527.16 \text{ ft}$$

Find the station of the PT.

$$\begin{array}{r} \text{PC sta } 477{+}38.50 \\ +\ 15\ \ 27.16 \\ \hline \text{PT sta } 492{+}65.66 \end{array} \quad (\text{sta } 492{+}65)$$

The answer is (C).

Why Other Options Are Wrong

(A) In this incorrect answer, using the full deflection angle instead of dividing by 2 in the tangent length equation will produce the wrong point of curvature (PC) station.

(B) In this incorrect answer, using the formula for the chord length and adding that dimension to the point of curvature (PC) station provides a station that is too low, since the stationing will not follow the centerline.

(D) In this incorrect answer, adding the tangent length to the point of intersection (PI) provides a possible second station of the point of tangent (PT), but this is not correct practice.

SOLUTION 16

As recommended by the GDHS, Formula 3-33 is used for determining the maximum comfortable speed on horizontal curves.

$$\frac{e}{100} + f_{\max} = \frac{v^2}{15R}$$

Rearranging to find R,

$$R = \frac{v^2}{15\left(\frac{e}{100} + f_{\max}\right)}$$

The maximum friction factor, $f_{\max}$, is shown in Exh. 3-41 to be 0.221 for a design speed of 30 mph.

Solving for the radius,

$$R = \frac{\left(30\ \frac{\text{mi}}{\text{hr}}\right)^2}{\left(15\ \frac{\text{mi}^2}{\text{hr}^2\text{-ft}}\right)\left(\frac{-4\%}{100} + 0.221\right)}$$
$$= 331 \text{ ft} \quad (330 \text{ ft})$$

The answer is (D).

Why Other Options Are Wrong

(A) This incorrect answer results from entering the cross slope directly in percent without converting to feet per foot and then ignoring the negative sign in the answer.

(B) This incorrect answer results from ignoring the negative sign for the cross slope.

(C) This incorrect answer results from using Exh. 3-21 of the GDHS. This exhibit is based on a positive cross slope (i.e., a scenario where the outer edge of a turning roadway is high) instead of on a negative cross slope (which is when the outer edge is lower than the inner edge).

SOLUTION 17

The rate of change of grade is uniform along the curve. The point where the grade is zero will be proportional to the distance along the curve related to the grade change for the total curve length.

$$x = \frac{G_1 L}{G_1 - G_2} = \frac{(-4\%)(6 \text{ sta})}{-4\% - 5\%}$$
$$= \frac{-24 \text{ sta-}\%}{-9\%}$$
$$= 2.667 \text{ sta} \quad (266.70 \text{ ft})$$

The low point is 266.70 ft ahead from the PVC.

The answer is (D).

Why Other Options Are Wrong

(A) In this incorrect answer, the proportional distance along the curve is incomplete because it was not multiplied by the curve length.

(B) This incorrect answer is the station distance, misconstrued as feet.

(C) In this incorrect answer, the proportional distance along the curve was not multiplied by the curve length and was converted to feet.

SOLUTION 18

Find the elevation of the point of vertical curve (PVC).

$$E_{\text{PVC}} = E_{\text{PVI}} + G_1\left(\frac{\text{LVC}}{2}\right)$$
$$= 425.38 \text{ ft} + \left(3.6\ \frac{\text{ft}}{\text{sta}}\right)(2.5 \text{ sta})$$
$$= 434.38 \text{ ft}$$

Find the rate of change of grade per station.

$$R = \frac{G_2 - G_1}{L} = \frac{4.4\% - (-3.6\%)}{5 \text{ sta}}$$
$$= 1.60\%/\text{sta}$$

Use that rate to determine the elevation of the point under the overpass. The elevation of the PVC is used as the known point in the equation.

$$E_P = E_{PVC} + G_1 x + \left(\frac{R}{2}\right) x^2$$

$$= 434.38 \text{ ft}$$
$$+ \begin{pmatrix} (-3.6\%)(82.45 \text{ sta} - 81.00 \text{ sta}) \\ + \left(\frac{1.6\%}{2 \text{ sta}}\right)(82.45 \text{ sta} - 81.00 \text{ sta})^2 \end{pmatrix}$$
$$\times \left(100 \frac{\text{ft}}{\text{sta}}\right)$$
$$= 430.84 \text{ ft} \quad (431 \text{ ft})$$

The answer is (C).

Why Other Options Are Wrong

(A) In this incorrect answer, the elevation of the PVC was incorrectly determined.

$$E_{PVC} = E_{PVI} + g_1 \left(\frac{LVC}{2}\right)$$
$$= 425.38 \text{ ft} + \left(-3.6 \frac{\text{ft}}{\text{sta}}\right)(2.5 \text{ sta})$$
$$= 416.38 \text{ ft}$$

$$E_P = E_{PVC} + G_1 x - \left(\frac{R}{2}\right) x^2$$
$$= 416.38 \text{ ft} - 5.22 \text{ ft} + 1.68 \text{ ft}$$
$$= 412.84 \text{ ft} \quad (413 \text{ ft})$$

(B) In this incorrect answer, the distance term was not squared.

$$E_P = E_{PVC} + G_1 x + \left(\frac{R}{2}\right) x^2$$
$$= 434.38 \text{ ft} + \begin{pmatrix} (-3.6\%)(82.45 \text{ sta} - 81.00 \text{ sta}) \\ + \left(\frac{1.6\%}{2 \text{ sta}}\right)(82.45 \text{ sta} - 81.00 \text{ sta}) \end{pmatrix}$$
$$\times \left(100 \frac{\text{ft}}{\text{sta}}\right)$$
$$= 430.32 \text{ ft} \quad (430 \text{ ft})$$

(D) In this incorrect answer, in the second term of the elevation equation, the negative sign in front of the 3.6% grade was missed.

SOLUTION 19

Determine the distance between the two known points.

$$x = 8150 \text{ ft} - 7600 \text{ ft} = 550 \text{ ft}$$

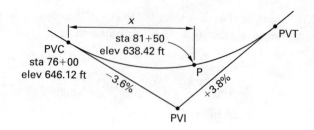

The rate of change can be related to the unknown length of vertical curve.

$$R = \frac{G_2 - G_1}{L} = \frac{0.038 \frac{\text{ft}}{\text{ft}} - \left(-0.036 \frac{\text{ft}}{\text{ft}}\right)}{L}$$
$$= \frac{0.074 \frac{\text{ft}}{\text{ft}}}{L}$$

$$\frac{R}{2} = \frac{0.037 \frac{\text{ft}}{\text{ft}}}{L}$$

Insert values into the complete elevation formula.

$$E_P = E_{PVC} + G_1 x + \left(\frac{R}{2}\right) x^2$$

$$638.42 \text{ ft} = 646.12 \text{ ft} + \left(-0.036 \frac{\text{ft}}{\text{ft}}\right)(550 \text{ ft})$$
$$+ \left(\frac{0.037 \frac{\text{ft}}{\text{ft}}}{L}\right)(550 \text{ ft})^2$$

Rearranging,

$$638.42 \text{ ft} - 646.12 \text{ ft} + 19.8 \text{ ft} = \frac{\left(0.037 \frac{\text{ft}}{\text{ft}}\right)(550 \text{ ft})^2}{L}$$

$$L = \frac{\left(0.037 \frac{\text{ft}}{\text{ft}}\right)(550 \text{ ft})^2}{12.1 \text{ ft}}$$
$$= 925 \text{ ft}$$

The answer is (D).

Why Other Options Are Wrong

(A) Incorrectly determining the difference in absolute grades instead of the total change in grade results in a very short grade length.

(B) This answer is incorrect. Neglecting to square the distance in the third term and assuming the answer is in stations because it is so small results in a vertical curve that is much too short.

$$638.42 \text{ ft} = 646.12 \text{ ft} + \left(-0.036 \frac{\text{ft}}{\text{ft}}\right)(550 \text{ ft})$$
$$+ \left(\frac{0.037 \frac{\text{ft}}{\text{ft}}}{L}\right)(550 \text{ ft})$$

Rearranging,

$$638.42 \text{ ft} - 646.12 \text{ ft} + 19.8 \text{ ft} = \frac{\left(0.037 \frac{\text{ft}}{\text{ft}}\right)(550 \text{ ft})}{L}$$

$$L = \frac{\left(0.037 \frac{\text{ft}}{\text{ft}}\right)(550 \text{ ft})}{12.1 \text{ ft}}$$
$$= 1.68 \text{ sta} \quad (168 \text{ ft})$$

(C) This answer is incorrect. Omitting the minus sign for G_1 in the second term of the elevation equation and then ignoring the minus sign of the answer results in a curve that is too short.

SOLUTION 20

Convert travel speed to feet per second.

$$v_{\text{horiz}} = \left(150 \frac{\text{mi}}{\text{hr}}\right)\left(5280 \frac{\text{ft}}{\text{mi}}\right)\left(\frac{1 \text{ hr}}{3600 \text{ sec}}\right)$$
$$= 220 \text{ ft/sec}$$

Find the vertical acceleration limit using $1g = 32.2 \text{ fps}^2$.

$$a_{\text{vert}} = (0.03)\left(32.2 \frac{\text{ft}}{\text{sec}^2}\right)$$
$$= 0.966 \text{ ft/sec}^2$$

When the train is traveling on a downgrade, it has a negative vertical velocity. On an upgrade, the train has a positive vertical velocity.

Vertical velocity on the downgrade is

$$v_{\text{vert}} = G_1 S = \left(\frac{-0.75 \text{ ft}}{100 \text{ ft}}\right)\left(220 \frac{\text{ft}}{\text{sec}}\right)$$
$$= -1.650 \text{ ft/sec}$$

Vertical velocity on the upgrade is

$$v_{\text{vert}} = G_2 S = \left(\frac{0.45 \text{ ft}}{100 \text{ ft}}\right)\left(220 \frac{\text{ft}}{\text{sec}}\right)$$
$$= 0.990 \text{ ft/sec}$$

Determine the total change in vertical velocity.

$$\Delta v = |v_{\text{down}} - v_{\text{up}}|$$
$$= \left|\left(-1.650 \frac{\text{ft}}{\text{sec}}\right) - \left(0.990 \frac{\text{ft}}{\text{sec}}\right)\right|$$
$$= 2.64 \text{ ft/sec}$$

Determine the number of seconds required to change direction.

$$t = \frac{\Delta v}{a_{\text{vert}}} = \frac{2.64 \frac{\text{ft}}{\text{sec}}}{0.966 \frac{\text{ft}}{\text{sec}^2}}$$
$$= 2.73 \text{ sec}$$

The distance traveled in 2.73 sec is the minimum length of vertical curve required.

$$L_{\min} = t v_{\text{horiz}} = (2.73 \text{ sec})\left(220 \frac{\text{ft}}{\text{sec}}\right)$$
$$= 601 \text{ ft} \quad (600 \text{ ft})$$

The answer is (C).

Why Other Options Are Wrong

(A) In this incorrect answer, using the difference in absolute vertical velocity values results in too short of a curve.

(B) In this incorrect answer, neglecting to convert miles per hour into feet per second results in too short of a curve.

(D) In this incorrect answer, inserting the metric value for gravity acceleration and not converting to feet per second squared to determine the vertical acceleration limit results in too long of a curve.

$$a_{\text{vert}} = (0.03)\left(9.807 \frac{\text{ft}}{\text{sec}^2}\right)$$
$$= 0.294 \text{ ft/sec}^2$$

Determine number of seconds required to change direction.

$$t = \frac{\Delta v}{a_{\text{vert}}} = \frac{2.64 \frac{\text{ft}}{\text{sec}}}{0.294 \frac{\text{ft}}{\text{sec}^2}}$$
$$= 8.97 \text{ sec}$$

Distance traveled in 2.73 sec is the minimum length of vertical curve required.

$$L_{\min} = t v_{\text{horiz}} = (8.97 \text{ sec})\left(220 \frac{\text{ft}}{\text{sec}}\right)$$
$$= 1974 \text{ ft} \quad (1980 \text{ ft})$$

Depth Solutions

TRAFFIC ANALYSIS

SOLUTION 21

Use the base relationship

$$D = \frac{\text{unit length}}{\text{bicycle spacing}}$$

Determine the following.

$$\begin{aligned}
V &= (\text{directional factor})(\text{two-way flow}) \\
&= (0.6)\left(\frac{20 \text{ bicycles}}{15 \text{ min}}\right)\left(60 \frac{\text{min}}{\text{hr}}\right) \\
&= 48 \text{ bicycles/hr}
\end{aligned}$$

$$\begin{aligned}
h &= \frac{\text{time unit}}{\text{arrivals during time unit}} \\
&= \frac{60 \frac{\text{min}}{\text{hr}}}{48 \frac{\text{bicycles}}{\text{hr}}} \\
&= 1.25 \text{ min/bicycle}
\end{aligned}$$

$$\begin{aligned}
S &= \left(11.2 \frac{\text{mi}}{\text{hr}}\right)\left(5280 \frac{\text{ft}}{\text{mi}}\right)\left(\frac{1 \text{ hr}}{3600 \text{ sec}}\right) \\
&= 16.4 \text{ ft/sec}
\end{aligned}$$

$$\begin{aligned}
\text{bicycle spacing} &= hS \\
&= \left(1.25 \frac{\text{min}}{\text{bicycle}}\right)\left(16.4 \frac{\text{ft}}{\text{sec}}\right)\left(60 \frac{\text{sec}}{\text{min}}\right) \\
&= 1230 \text{ ft/bicycle}
\end{aligned}$$

$$\begin{aligned}
D &= \frac{5280 \frac{\text{ft}}{\text{mi}}}{1230 \frac{\text{ft}}{\text{bicycle}}} \\
&= 4.29 \text{ bicycles/mi} \quad (4.3 \text{ bicycles/mi})
\end{aligned}$$

The answer is (B).

Alternate Solution:

$$D = \frac{V}{S} = \frac{48 \frac{\text{bicycles}}{\text{hr}}}{11.2 \frac{\text{mi}}{\text{hr}}} = 4.3 \text{ bicycles/mi}$$

Why Other Options Are Wrong

(A) This incorrect answer results from inverting the density determination.

(C) This incorrect answer results from using the two-directional flow as the one-way flow.

(D) This incorrect answer results from incorrectly applying the density and flow rate relationship.

SOLUTION 22

A common unit of vehicle measure is necessary to make traffic evaluations; therefore, passenger-car equivalents, option I, is chosen as the standard of reference.

The length of the standard reference vehicle, option IV, determines how much roadway surface is occupied by vehicles.

Vehicle spacing, option VI, determines how many vehicles occupy a unit length of roadway, such as one lane-mile, at an instant of time.

The answer is (A).

Why Other Options Are Wrong

(B) This answer is incorrect. Vehicle count data, option III, must be converted to passenger-car equivalents.

(C) This answer is incorrect. Vehicle count data, option III, must be converted to passenger-car equivalents, and the weight of vehicles, option V, is not considered in problems addressing traffic density.

(D) This answer is incorrect. Density, as a definition of vehicle occupancy of a roadway lane, is independent of speed, so option II does not apply.

SOLUTION 23

The service flow rate is the hourly equivalent rate that flows in the peak 15 min.

Passenger-car equivalents are determined by the heavy vehicle factor, option II, and the driver population factor, option I.

The number of lanes included in the vehicle count, option IV, must be known so that the count can be averaged to a single lane.

The peak-hour factor, option VI, is used to determine how much of the peak-hour traffic flows in the peak 15 min.

The answer is (A).

Why Other Options Are Wrong

(B) This answer is incorrect. The volume-over-capacity ratio, option VIII, is determined after the service flow rate is determined and is not a direct factor in the service flow rate determination.

(C) This answer is incorrect. Lane width and parking adjustment factors, options III and V, affect service volume determinations for intersection approaches, not service flow rates for highways.

(D) This answer is incorrect. Lane width and parking adjustment factors, options III and V, affect service volume determinations for intersection approaches, not service flow rates for highways. Average control delay per vehicle, option VII, applies to controlled intersections.

SOLUTION 24

Examining HCM Exh. 23-9 shows that, as the percentage of trucks increases beyond 2%, the passenger-car equivalence reduces, particularly on longer grades.

The answer is (B).

Why Other Options Are Wrong

(A) This answer is incorrect. For general freeway sections, the most common traffic pattern shows a morning peak flow and an evening peak flow, which is attributable to auto commuters. Truck traffic, on the other hand, tends to increase during mid-day near urban areas because of local delivery activity. Therefore, trucks are usually a lower percentage of peak-hour traffic than of off-peak traffic.

(C) This answer is incorrect. Trucks generally operate much slower than does other traffic on steep downgrades because of stopping safety. Therefore, trucks have a higher passenger-car equivalent on steep downgrades than they do on slight downgrades.

(D) This answer is incorrect. The passenger-car equivalent for RVs can be combined with the figure for trucks and buses when the percentage of trucks and buses is at least five times the percentage of RVs present.

SOLUTION 25

The HCM uses standardized periods of time for general traffic analysis of 1 hr intervals and 15 min intervals. The 15 min intervals of traffic are multiplied by four to determine 1 hr flow rates. By dividing a peak 1 hr flow rate by the peak 15 min flow rate within that hour, a comparison factor can be made to determine the uniformity of flow, called the peak-hour factor.

The answer is (A).

Why Other Options Are Wrong

(B) This answer is incorrect. While other analysis methods, such as peak within a peak, may use a 5 min flow rate, the HCM uses a 15 min flow rate for the peak-hour factor.

(C) This answer is incorrect. The peak 5 min flow rate divided by the peak 1 hr flow rate yields a simple flow fraction, but not a comparison of flow rates.

(D) This answer is incorrect. The peak 15 min flow rate divided by the peak 1 hr flow rate yields a simple flow fraction, but not a comparison of flow rates.

SOLUTION 26

Trucks descending long, steep downgrades generally travel slower than they would on long upgrades in order to avoid loss of braking power. Slower speeds on severe downgrades equate to larger passenger-car equivalents on downgrades than on upgrades. Statement (D) is false.

The answer is (D).

Why Other Options Are Wrong

(A) Buses in a traffic stream increase the passenger capacity far beyond the decrease in vehicle capacity based on auto equivalents. Statement (A) is true.

(B) Trucks require more auto-equivalent space and usually carry no more people, if not fewer, than autos. Therefore, the passenger capacity of the roadway is lower. Statement (B) is true.

(C) According to the HCM, RVs in small proportion to trucks can be included with trucks.

SOLUTION 27

The total queue to be dissipated is the sum of the 10 min vehicle accumulation plus the arrivals that continue while the queue is dissipating. The departure rate from the front of the queue is at the roadway capacity.

Determine the arrival rate.

$$\text{arrival rate} = \left(3100 \; \frac{\text{veh}}{\text{hr}}\right)\left(\frac{1 \; \text{hr}}{60 \; \text{min}}\right)$$
$$= 51.7 \; \text{vpm}$$

Determine the departure rate.

$$\text{departure rate} = \left(4200 \ \frac{\text{veh}}{\text{hr}}\right)\left(\frac{1 \ \text{hr}}{60 \ \text{min}}\right)$$
$$= 70 \ \text{vpm}$$

Equate the total vehicle accumulation to the departure rate, in relation to the unknown departure time, t.

$$\left(51.7 \ \frac{\text{veh}}{\text{min}}\right)(10 \ \text{min}) + \left(51.7 \ \frac{\text{veh}}{\text{min}}\right)t$$
$$= \left(70 \ \frac{\text{veh}}{\text{min}}\right)t$$

Solve for t.

$$\left(70 \ \frac{\text{veh}}{\text{min}}\right)t - \left(51.7 \ \frac{\text{veh}}{\text{min}}\right)t = 517 \ \text{veh}$$

$$\left(18.3 \ \frac{\text{veh}}{\text{min}}\right)t = 517 \ \text{veh}$$

$$t = \frac{517 \ \text{veh}}{18.3 \ \frac{\text{veh}}{\text{min}}} = 28.3 \ \text{min} \quad (28 \ \text{min})$$

This queue will be dissipated approximately 28 min after the freeway is reopened to full capacity.

The answer is (B).

Why Other Options Are Wrong

(A) This incorrect answer results from including in the calculation only the vehicles that arrive during the blockage.

(C) This incorrect answer results from including the 10 min blockage time in the departure time.

(D) This incorrect answer results from subtracting the 10 min blockage time from the departure time, t.

SOLUTION 28

The directional capacity is the capacity per lane times the number of lanes in each direction.

$$\text{directional capacity} = \left(\frac{\text{capacity}}{\text{hr-lane}}\right)(\text{no. of lanes})$$
$$= \left(2100 \ \frac{\text{veh}}{\text{hr-lane}}\right)(2 \ \text{lanes})$$
$$= 4200 \ \text{vph}$$

The arrival rate is more than the one-lane capacity. Therefore, when one lane is blocked there will be a queue of vehicles behind the blockage. Once the blocked lane is opened, the capacity will be greater than the arrival rate, so the queue will begin to dissipate. The total queue to be dissipated is as follows.

$$\text{total queue} = \sum \text{arrivals} - \sum \text{departures}$$
$$= \text{arrivals during blockage}$$
$$+ \text{arrivals during queue dissipation}$$
$$- \text{departures during lane blockage}$$
$$- \text{departures during queue dissipation}$$

Arrivals for 20 min plus arrivals for time t equals departures at the one-lane rate for 20 min plus departures at the two-lane rate for time t.

Determine the arrival rate.

$$\text{arrival rate} = \left(3100 \ \frac{\text{veh}}{\text{hr}}\right)\left(\frac{1 \ \text{hr}}{60 \ \text{min}}\right)$$
$$= 51.7 \ \text{vpm}$$

Determine the two-lane departure rate, which is the full directional capacity.

$$\text{two-lane departure rate} = \left(4200 \ \frac{\text{veh}}{\text{hr}}\right)\left(\frac{1 \ \text{hr}}{60 \ \text{min}}\right)$$
$$= 70 \ \text{vpm}$$

Determine the one-lane departure rate, which is the departure rate during the accident blockage. (HCM exhibit 22-6.)

$$\text{accident blockage departure rate} = \left(70 \ \frac{\text{veh}}{\text{min}}\right)(0.35)$$
$$= 24.5 \ \text{vpm}$$

Time t is the time necessary to dissipate the queue. The time is 0 sec at the instant of clearing the blockage. Equate the total vehicle accumulation to the departure rate, in relation to the unknown departure time, t.

$$\text{total queue} = \left(51.7 \ \frac{\text{veh}}{\text{min}}\right)(20 \ \text{min}) + \left(51.7 \ \frac{\text{veh}}{\text{min}}\right)t$$
$$= \left(24.5 \ \frac{\text{veh}}{\text{min}}\right)(20 \ \text{min}) + \left(70 \ \frac{\text{veh}}{\text{min}}\right)t$$

Solve for t.

$$t = 29.7 \ \text{min} \quad (30 \ \text{min})$$

The entire queue will be dissipated in approximately 30 min.

The answer is (B).

Why Other Options Are Wrong

(A) This incorrect answer results from including in the calculation only the vehicle(s) that arrive during the blockage.

(C) This incorrect answer results from adding the 20 min blockage time to the queue dissipation time after the blockage.

(D) This incorrect answer results from not subtracting the single-lane departures during the 20 min lane blockage.

SOLUTION 29

To determine the braking distance, s_b, the perception-reaction distance, s_r, must be subtracted from the total stopping distance, s_s.

$$s_b = s_s - s_r$$

$$s_r = \text{v}t$$

$$= \left(50 \ \frac{\text{mi}}{\text{hr}}\right)\left(5280 \ \frac{\text{ft}}{\text{mi}}\right)\left(\frac{1 \ \text{hr}}{3600 \ \text{sec}}\right)(2.5 \ \text{sec})$$

$$= 183.3 \ \text{ft}$$

$$s_b = 461 \ \text{ft} - 183.3 \ \text{ft}$$

$$= 277.7 \ \text{ft}$$

The friction factor, f, is found from the formula for deceleration.

$$s_b = \frac{\text{v}_1^2 - \text{v}_2^2}{2(f+G)}$$

Rearrange to find f, setting v_2 at 0 mph, G at 0 ft/ft, and g at 32.2 ft/sec^2.

$$f = \frac{\text{v}_1^2}{2gs_b}$$

$$= \frac{\left(\left(50 \ \frac{\text{mi}}{\text{hr}}\right)\left(5280 \ \frac{\text{ft}}{\text{mi}}\right)\left(\frac{1 \ \text{hr}}{3600 \ \text{sec}}\right)\right)^2}{(2)\left(32.2 \ \frac{\text{ft}}{\text{sec}^2}\right)(277.7 \ \text{ft})}$$

$$= 0.30$$

Determine the braking distance from 60 mph using the same friction factor.

$$s_{b,60} = \frac{\text{v}_1^2 - \text{v}_2^2}{2(f+G)}$$

$$= \frac{\left(\left(60 \ \frac{\text{mi}}{\text{hr}}\right)\left(5280 \ \frac{\text{ft}}{\text{mi}}\right)\left(\frac{1 \ \text{hr}}{3600 \ \text{sec}}\right)\right)^2 - \left(0 \ \frac{\text{mi}}{\text{hr}}\right)^2}{(2)\left(32.2 \ \frac{\text{ft}}{\text{sec}^2}\right)\left(0.30 + 0 \ \frac{\text{ft}}{\text{ft}}\right)}$$

$$= 400.8 \ \text{ft}$$

The perception-reaction distance for 60 mph is

$$s_{r,60} = s_{s,60} - s_{b,60}$$

$$= 461 \ \text{ft} - 400.8 \ \text{ft}$$

$$= 60.2 \ \text{ft}$$

The perception-reaction time required at 60 mph is

$$t_{r,60} = \frac{s_r}{\text{v}}$$

$$= \frac{60.2 \ \text{ft}}{\left(60 \ \frac{\text{mi}}{\text{hr}}\right)\left(5280 \ \frac{\text{ft}}{\text{mi}}\right)\left(\frac{1 \ \text{hr}}{3600 \ \text{sec}}\right)}$$

$$= 0.68 \ \text{sec} \quad (0.7 \ \text{sec})$$

The answer is (C).

Why Other Options Are Wrong

(A) This incorrect answer results from an improper conversion from miles per hour to feet per second.

(B) This incorrect answer results from an improper conversion from miles per hour to feet per second, ignoring the negative value and assuming the answer must be a positive value.

(D) This incorrect answer results from neglecting the perception-reaction distance at 50 mph to establish the braking friction factor, thereby assuming an average friction factor for the entire stopping distance. The answer assumes a positive value.

SOLUTION 30

The lead car stops in 278 ft, which is the braking distance, $s_{b,\text{lead}}$. The braking distance determines the minimum friction factor, f.

$$s_{b,\text{lead}} = \frac{\text{v}_1^2 - \text{v}_2^2}{2g(f+G)}$$

Setting v_2 at 0 mph and g at 32.2 ft/sec^2 and solving for f,

$$f = \frac{\text{v}_1^2}{2gs_{b,50}}$$

$$= \frac{\left(\left(50 \ \frac{\text{mi}}{\text{hr}}\right)\left(5280 \ \frac{\text{ft}}{\text{mi}}\right)\left(\frac{1 \ \text{hr}}{3600 \ \text{sec}}\right)\right)^2}{(2)\left(32.2 \ \frac{\text{ft}}{\text{sec}^2}\right)(278 \ \text{ft})}$$

$$= 0.30$$

The stopping distance, s_s, for the following car includes braking distance, s_b, and perception-reaction distance, s_r.

$$s_s = s_r + s_b$$

Find the braking distance from 60 mph using the friction factor for the lead car.

$$s_{b,60} = \frac{v_1^2 - v_2^2}{2g(f+G)}$$

$$= \frac{\left(\left(60 \frac{\text{mi}}{\text{hr}}\right)\left(5280 \frac{\text{ft}}{\text{mi}}\right)\left(\frac{1 \text{ hr}}{3600 \text{ sec}}\right)\right)^2 - \left(0 \frac{\text{mi}}{\text{hr}}\right)^2}{(2)\left(32.2 \frac{\text{ft}}{\text{sec}^2}\right)\left(0.30 + 0.0 \frac{\text{ft}}{\text{ft}}\right)}$$

$$= 400.8 \text{ ft}$$

The perception-reaction distance for the following car is

$$s_{r,60} = vt_r$$
$$= \left(60 \frac{\text{mi}}{\text{hr}}\right)\left(5280 \frac{\text{ft}}{\text{mi}}\right)\left(\frac{1 \text{ hr}}{3600 \text{ sec}}\right)(2.0 \text{ sec})$$
$$= 176 \text{ ft}$$

The stopping distance from 60 mph, $s_{s,60}$, is

$$s_{s,60} = s_{r,60} + s_{b,60}$$
$$= 176 \text{ ft} + 400.8 \text{ ft}$$
$$= 576.8 \text{ ft}$$

The minimum following distance is

$$s_f = s_{s,60} - s_{b,50}$$
$$= 576.8 \text{ ft} - 278 \text{ ft}$$
$$= 298.8 \text{ ft} \quad (300 \text{ ft})$$

The answer is (C).

Why Other Options Are Wrong

(A) This incorrect answer results from assuming the braking distance for the lead car includes perception-reaction time (2.0 sec).

(B) This answer results from incorrectly subtracting the length of the lead car (20 ft standard from the GDHS) from the available stopping distance for the following car.

(D) This incorrect answer results from adding the length of the lead car (20 ft standard from the GDHS) to the stopping distance available for the following car.

SOLUTION 31

The total distance between stations has three components of travel.

$$s_{\text{total}} = s_{\text{accl}} + s_{\text{running}} + s_{\text{decel}} = 5280 \text{ ft}$$

Determine the distance necessary to accelerate to 80 mph and the distance necessary to decelerate from 80 mph to a stop.

$$s_{\text{accel}} = \frac{v^2 - v_o^2}{2a}$$

$$= \left(\frac{\left(80 \frac{\text{mi}}{\text{hr}}\right)^2 - \left(0 \frac{\text{mi}}{\text{hr}}\right)^2}{(2)\left(5.5 \frac{\text{ft}}{\text{sec}^2}\right)}\right)\left(5280 \frac{\text{ft}}{\text{mi}}\right)^2$$

$$\times \left(\frac{1 \text{ hr}}{3600 \text{ sec}}\right)^2$$

$$= 1252 \text{ ft}$$

$$s_{\text{decel}} = \frac{v^2 - v_o^2}{2d}$$

$$= \left(\frac{\left(0 \frac{\text{mi}}{\text{hr}}\right)^2 - \left(80 \frac{\text{mi}}{\text{hr}}\right)^2}{(2)\left(-4.4 \frac{\text{ft}}{\text{sec}^2}\right)}\right)\left(5280 \frac{\text{ft}}{\text{mi}}\right)^2$$

$$\times \left(\frac{1 \text{ hr}}{3600 \text{ sec}}\right)^2$$

$$= 1564 \text{ ft}$$

Subtract the acceleration and deceleration distances from 1 mi to find the distance of constant running at 80 mph.

$$d_{80 \text{ mph}} = 5280 \text{ ft} - s_{\text{accel}} - s_{\text{decel}}$$
$$= 5280 \text{ ft} - 1252 \text{ ft} - 1564 \text{ ft}$$
$$= 2464 \text{ ft}$$

Determine the time required for the three components of travel.

$$t_{\text{accel}} = \frac{v - v_o}{a}$$

$$= \left(\frac{80 \frac{\text{mi}}{\text{hr}} - 0 \frac{\text{mi}}{\text{hr}}}{5.5 \frac{\text{ft}}{\text{sec}^2}}\right)\left(5280 \frac{\text{ft}}{\text{mi}}\right)\left(\frac{1 \text{ hr}}{3600 \text{ sec}}\right)$$

$$= 21.3 \text{ sec}$$

$$t_{\text{decel}} = \frac{v - v_o}{d}$$

$$= \left(\frac{0 \frac{\text{mi}}{\text{hr}} - 80 \frac{\text{mi}}{\text{hr}}}{-4.4 \frac{\text{ft}}{\text{sec}^2}}\right)\left(5280 \frac{\text{ft}}{\text{mi}}\right)\left(\frac{1 \text{ hr}}{3600 \text{ sec}}\right)$$

$$= 26.7 \text{ sec}$$

$$t_{80\text{ mph}} = \frac{s}{v}$$

$$= \left(\frac{2464 \text{ ft}}{80 \frac{\text{mi}}{\text{hr}}}\right)\left(\frac{1 \text{ mi}}{5280 \text{ ft}}\right)\left(3600 \frac{\text{sec}}{\text{hr}}\right)$$

$$= 21.0 \text{ sec}$$

The total time to travel between stations is the sum of the component times.

$$t_{\text{total}} = t_{\text{accel}} + t_{\text{decel}} + t_{80\text{ mph}}$$
$$= 21.3 \text{ sec} + 26.7 \text{ sec} + 21.0 \text{ sec}$$
$$= 69.0 \text{ sec}$$

Determine the average travel speed between stations.

$$v = \left(\frac{1 \text{ mi}}{69.0 \text{ sec}}\right)\left(3600 \frac{\text{sec}}{\text{hr}}\right)$$
$$= 52.2 \text{ mph} \quad (52 \text{ mph})$$

The answer is (C).

Why Other Options Are Wrong

(A) In this incorrect answer, using the 5 min train schedule spacing as the time required to travel from one station to the next gives a result that is too small.

(B) This incorrect answer results from using 80 mph as the top instant speed, ignoring the 80 mph constant running time and proportioning the acceleration and deceleration rates to determine the acceleration and deceleration times covering the entire 1 mi distance between stations.

$$s_{\text{accel}} = \frac{v^2 - v_o^2}{2a} = \left(\frac{4.4}{5.5 + 4.4}\right)(5280 \text{ ft})$$
$$= 2347 \text{ ft}$$

$$s_{\text{decel}} = \frac{v^2 - v_o^2}{2d} = 5280 \text{ ft} - 2347 \text{ ft}$$
$$= 2933 \text{ ft}$$

$$t_{\text{accel}} = \frac{2s_{\text{accel}}}{v_o + v}$$

$$= \left(\frac{(2)(2347 \text{ ft})}{0 \frac{\text{mi}}{\text{hr}} + 80 \frac{\text{mi}}{\text{hr}}}\right)\left(\frac{1 \text{ mi}}{5280 \text{ ft}}\right)\left(3600 \frac{\text{sec}}{\text{hr}}\right)$$
$$= 40 \text{ sec}$$

$$t_{\text{decel}} = \frac{2s_{\text{decel}}}{v_o - v}$$

$$= \left(\frac{(2)(2933 \text{ ft})}{80 \frac{\text{mi}}{\text{hr}}}\right)\left(\frac{1 \text{ mi}}{5280 \text{ ft}}\right)\left(3600 \frac{\text{sec}}{\text{hr}}\right)$$
$$= 50 \text{ sec}$$

Because 80 mph is used as the instant top speed, there is no running time at constant speed. The total time between stations is the sum of the acceleration and deceleration times.

$$t_{\text{total}} = t_{\text{accel}} + t_{\text{decel}} = 40 \text{ sec} + 50 \text{ sec}$$
$$= 90 \text{ sec}$$

Determine the average speed.

$$v = \left(\frac{1 \text{ mi}}{90 \text{ sec}}\right)\left(3600 \frac{\text{sec}}{\text{hr}}\right) = 40 \text{ mph}$$

(D) This incorrect answer results from failing to convert from miles per hour to feet per second.

SOLUTION 32

Determine the total number of train cars arriving per hour.

$$\text{total trains per hour} = \left(\frac{1}{\text{train spacing}}\right)$$
$$= \left(\frac{1 \text{ train}}{5 \text{ min}}\right)\left(60 \frac{\text{min}}{\text{hr}}\right)$$
$$= 12 \text{ trains/hr}$$

$$\text{no. of cars per hour} = (\text{no. of trains per hour})$$
$$\times (\text{no. of cars per train})$$
$$= \left(4 \frac{\text{cars}}{\text{train}}\right)\left(12 \frac{\text{trains}}{\text{hr}}\right)$$
$$= 48 \text{ cars/hr}$$

Multiply the number of cars per hour by the capacity of each car, to determine the number of passengers per hour.

$$\left(48 \frac{\text{cars}}{\text{hr}}\right)\left(220 \frac{\text{passengers}}{\text{car}}\right) = 10{,}560 \text{ passenger/hr}$$
$$(10{,}600 \text{ passengers/hr})$$

The answer is (C).

Why Other Options Are Wrong

(A) This incorrect answer results from not including the number of trains in each hour.

(B) This incorrect answer results from not including the number of cars in each train.

(D) This incorrect answer results from subtracting the dwell time from the arrival time, which is the time between trains, and then using the time between trains to determine the number of trains per hour.

$$\text{time between trains} = 5 \text{ min} - 1 \text{ min}$$
$$= 4 \text{ min}$$

$$\left(60 \frac{\text{min}}{\text{hr}}\right)\left(\frac{1 \text{ train}}{4 \text{ min}}\right) = 15 \text{ trains/hr}$$

DEPTH SOLUTIONS **41**

$$\left(15 \ \frac{\text{trains}}{\text{hr}}\right)\left(4 \ \frac{\text{cars}}{\text{train}}\right)\left(220 \ \frac{\text{passengers}}{\text{car}}\right)$$
$$= 13{,}200 \ \text{passengers/hr}$$

SOLUTION 33

The total distance between stations includes acceleration distance and deceleration distance only.

$$s_{\text{total}} = s_{\text{accel}} + s_{\text{decel}}$$

Determine the acceleration distance to 150 mph using the given acceleration rate.

$$s_{\text{accel}} = \frac{v_2^2 - v_1^2}{2a}$$
$$= \frac{\left(\left(150 \ \frac{\text{mi}}{\text{hr}}\right)\left(5280 \ \frac{\text{ft}}{\text{mi}}\right)\left(\frac{1 \ \text{hr}}{3600 \ \text{sec}}\right)\right)^2 - \left(0 \ \frac{\text{mi}}{\text{hr}}\right)^2}{(2)(0.18)\left(32.2 \ \frac{\text{ft}}{\text{sec}^2}\right)}$$
$$= 4175 \ \text{ft}$$

Determine the deceleration distance from 150 mph using the given deceleration rate.

$$s_{\text{decel}} = \frac{v_1^2 - v_2^2}{2d}$$
$$= \frac{\left(0 \ \frac{\text{mi}}{\text{hr}}\right)^2 - \left(\left(150 \ \frac{\text{mi}}{\text{hr}}\right)\left(5280 \ \frac{\text{ft}}{\text{mi}}\right)\left(\frac{1 \ \text{hr}}{3600 \ \text{sec}}\right)\right)^2}{(2)(-0.12)\left(32.2 \ \frac{\text{ft}}{\text{sec}^2}\right)}$$
$$= 6263 \ \text{ft}$$

Determine the total distance to accelerate to 150 mph then decelerate to a stop.

$$s_{\text{total}} = (4175 \ \text{ft} + 6263 \ \text{ft})\left(\frac{1 \ \text{mi}}{5280 \ \text{ft}}\right)$$
$$= 1.98 \ \text{mi} \quad (2.0 \ \text{mi})$$

The minimum station spacing is approximately 2.0 mi.

The answer is (C).

Why Other Options Are Wrong

(A) This incorrect distance covers acceleration to 150 mph only.

(B) This answer results from an incorrect conversion from miles per hour to feet per second.

(D) This incorrect answer results from using the metric value of gravity acceleration without converting to feet per second.

SOLUTION 34

Travel between stations includes three conditions of motion: acceleration, travel at constant speed, and deceleration. Travel at constant speed uses the distance remaining between acceleration and deceleration distances.

$$s_{\text{total}} = s_{\text{accel}} + s_{\text{running}} + s_{\text{decel}}$$

Determine the acceleration distance.

$$s_{\text{accel}} = \frac{v_2^2 - v_1^2}{2a}$$
$$= \frac{\left(\left(80 \ \frac{\text{mi}}{\text{hr}}\right)\left(5280 \ \frac{\text{ft}}{\text{mi}}\right)\left(\frac{1 \ \text{hr}}{3600 \ \text{sec}}\right)\right)^2 - \left(0 \ \frac{\text{mi}}{\text{hr}}\right)^2}{(2)\left(5.5 \ \frac{\text{ft}}{\text{sec}^2}\right)}$$
$$= 1252 \ \text{ft}$$

Determine the deceleration distance.

$$s_{\text{decel}} = \frac{v_2^2 - v_1^2}{2d}$$
$$= \frac{\left(0 \ \frac{\text{mi}}{\text{hr}}\right)^2 - \left(\left(80 \ \frac{\text{mi}}{\text{hr}}\right)\left(5280 \ \frac{\text{ft}}{\text{mi}}\right)\left(\frac{1 \ \text{hr}}{3600 \ \text{sec}}\right)\right)^2}{(2)\left(-4.5 \ \frac{\text{ft}}{\text{sec}^2}\right)}$$
$$= 1530 \ \text{ft}$$

Find the distance available for constant running speed.

$$s_{\text{running}} = s_{\text{total}} - s_{\text{accel}} - s_{\text{decel}}$$
$$= (1 \ \text{mi})\left(5280 \ \frac{\text{ft}}{\text{mi}}\right) - 1252 \ \text{ft} - 1530 \ \text{ft}$$
$$= 2498 \ \text{ft}$$

The total travel time is the sum of acceleration time, constant-speed running time, and deceleration time.

$$t_{\text{total}} = t_{\text{accel}} + t_{\text{running}} + t_{\text{decel}}$$

Determine acceleration time.

$$t_{accel} = \frac{v_2 - v_1}{a}$$

$$= \frac{\left(80 \frac{\text{mi}}{\text{hr}}\right)\left(5280 \frac{\text{ft}}{\text{mi}}\right)\left(\frac{1 \text{ hr}}{3600 \text{ sec}}\right) - 0 \frac{\text{mi}}{\text{hr}}}{5.5 \frac{\text{ft}}{\text{sec}^2}}$$

$$= 21.3 \text{ sec}$$

Determine constant running speed time.

$$t_{running} = \frac{s_{running}}{v}$$

$$= \frac{2498 \text{ ft}}{\left(80 \frac{\text{mi}}{\text{hr}}\right)\left(5280 \frac{\text{ft}}{\text{mi}}\right)\left(\frac{1 \text{ hr}}{3600 \text{ sec}}\right)}$$

$$= 21.3 \text{ sec}$$

Determine deceleration time.

$$t_{decel} = \frac{v_2 - v_1}{d}$$

$$= \frac{0 \frac{\text{mi}}{\text{hr}} - \left(80 \frac{\text{mi}}{\text{hr}}\right)\left(5280 \frac{\text{ft}}{\text{mi}}\right)\left(\frac{1 \text{ hr}}{3600 \text{ sec}}\right)}{-4.5 \frac{\text{ft}}{\text{sec}^2}}$$

$$= 26.1 \text{ sec}$$

Determine total time.

$$t_{total} = 21.3 \text{ sec} + 21.3 \text{ sec} + 26.1 \text{ sec}$$
$$= 68.7 \text{ sec}$$

The average running speed is the sum of the distances traveled divided by the sum of the times traveled. Determine the average running speed.

$$v_{ave} = \frac{s_{total}}{t_{total}}$$

$$= \left(\frac{5280 \text{ ft}}{68.7 \text{ sec}}\right)\left(\frac{1 \text{ mi}}{5280 \text{ ft}}\right)\left(3600 \frac{\text{sec}}{\text{hr}}\right)$$

$$= 52.4 \text{ mph} \quad (52 \text{ mph})$$

The answer is (C).

Why Other Options Are Wrong

(A) This answer results from incorrectly including the dwell time at both stations in the total time between two stations.

(B) This answer results from incorrectly including the dwell time at one station in the total time between stations.

(D) This answer results from not converting miles per hour to feet per second.

SOLUTION 35

The average speed is the total running time plus the delay per stop divided into the 10 mi length of trip.

$$v_{ave} = \frac{s_{total}}{t_{running} + (30 \text{ sec})(\text{number of stops})}$$

Determine the running time without stopping.

$$t = \frac{s_{total}}{v_{running}} = \frac{10 \text{ mi}}{\left(35 \frac{\text{mi}}{\text{hr}}\right)\left(\frac{1 \text{ hr}}{60 \text{ min}}\right)}$$

$$= 17.1 \text{ min}$$

Add the delay time for the current number of stops.

$$t_{delay} = \left(\begin{array}{c}\text{no. of stops} \\ \text{per mile}\end{array}\right)\left(\begin{array}{c}\text{length of} \\ \text{line}\end{array}\right)\left(\begin{array}{c}\text{delay} \\ \text{per stop}\end{array}\right)$$

$$= \left(4 \frac{\text{stops}}{\text{mi}}\right)(10 \text{ mi})\left(30 \frac{\text{sec}}{\text{stop}}\right)\left(\frac{1 \text{ min}}{60 \text{ sec}}\right)$$

$$= 20 \text{ min}$$

The current time for one trip is the running time plus the stop delay time.

$$t_{total} = t_{running} + t_{delay}$$
$$= 17.1 \text{ min} + 20 \text{ min} = 37.1 \text{ min}$$

Determine the current average speed.

$$v_{ave} = \frac{s_{total}}{t_{total}} = \left(\frac{10 \text{ mi}}{37.1 \text{ min}}\right)\left(60 \frac{\text{min}}{\text{hr}}\right)$$

$$= 16.2 \text{ mph}$$

Determine the delay time for the proposed number of stops.

$$t_{new\ delay} = \left(3 \frac{\text{stops}}{\text{mi}}\right)(10 \text{ mi})\left(30 \frac{\text{sec}}{\text{stop}}\right)\left(\frac{1 \text{ min}}{60 \text{ sec}}\right)$$

$$= 15 \text{ min}$$

Determine the proposed trip time.

$$t_{prop} = t_{running} + t_{new\ delay}$$
$$= 17.1 \text{ min} + 15 \text{ min}$$
$$= 32.1 \text{ min}$$

Determine the proposed average speed

$$v_{prop} = \frac{s}{t_{prop}} = \left(\frac{10 \text{ mi}}{32.1 \text{ min}}\right)\left(60 \frac{\text{min}}{\text{hr}}\right)$$

$$= 18.7 \text{ mph}$$

Determine the change in average speed.

$$\Delta v_{ave} = v_{prop} - v_{ave}$$
$$= 18.7 \,\frac{\text{mi}}{\text{hr}} - 16.2 \,\frac{\text{mi}}{\text{hr}}$$
$$= 2.5 \text{ mph}$$

The answer is (B).

Why Other Options Are Wrong

(A) This incorrect answer results from eliminating one deceleration and one acceleration from each mile, but not from the dwell time.

(C) This incorrect answer is the difference between the no-stop speed of 35 mph and the proposed speed for 3 stops/mi.

(D) This incorrect answer is the proposed average speed.

SOLUTION 36

The number of buses required is determined by dividing the number of bus riders (passengers) by the capacity of each bus.

$$\text{no. of buses} = \frac{\text{no. of bus riders}}{\text{capacity of each bus}}$$

The peak hour factor is the peak 15 min flow rate compared to the peak 1 hr flow rate. The arena will empty in 1 hour; therefore, the peak 1 hr flow is 12,000 persons. Determine how many people will ride buses, V, during one hour following the event.

$$V = (\% \text{ bus riders})(\text{people in attendance})$$
$$= (0.35)(12{,}000 \text{ persons})$$
$$= 4200 \text{ bus riders}$$

Define the peak hour factor (PHF), where V_{15} is the flow rate in persons per hour during the peak 15 min period.

$$\text{PHF} = \left(\frac{\text{flow volume in peak 1 hr}}{4(\text{flow volume in peak 15 min})}\right)$$
$$\times \left(\frac{V}{4V_{15}}\right)$$

Rearrange the definition to determine the number of bus riders during the 15 min peak.

$$V_{15} = \frac{V}{4(\text{PHF})} = \frac{4200 \text{ bus riders}}{(4)(0.85)}$$
$$= 1235 \text{ bus riders}$$

Determine the number of buses needed during the peak 15 min period.

$$\frac{\text{bus riders}}{\text{bus capacity}} = \frac{1235 \text{ bus riders}}{62 \,\frac{\text{bus riders}}{\text{bus}}}$$
$$= 19.9 \text{ buses} \quad (20 \text{ buses})$$

The answer is (B).

Why Other Options Are Wrong

(A) This incorrect answer assumes that the peak hour factor is applied to the total hour and then divided into 15 min periods.

(C) This incorrect answer projects the peak number of buses to an hourly demand.

(D) This incorrect answer divides the bus patronage into buses and then misapplies the peak hour factor to the resulting number of buses.

SOLUTION 37

Determine the UTF.

$$\text{UTF} = \left(\frac{1}{1000}\right)\left(\frac{\text{households}}{\text{auto}}\right)\left(\frac{\text{persons}}{\text{mi}^2}\right)$$
$$= \left(\frac{1}{1000}\right)\left(\frac{1 \text{ household}}{0.80 \text{ auto}}\right)\left(10{,}000 \,\frac{\text{persons}}{\text{mi}^2}\right)$$
$$= 12.50$$

Determine the percentage of trips on transit.

$$\% \text{ transit usage} = \frac{\text{UTF}}{0.6} = \frac{12.50}{0.6}$$
$$= 20.8\%$$

Determine the number of residents expected to use transit.

$$\text{no. of residents using transit} = (\text{transit usage})(\text{no. of residents})$$
$$= (0.208)(12{,}000)$$
$$= 2496 \quad (2500)$$

The answer is (D).

Why Other Options Are Wrong

(A) This incorrect answer results from applying the UTF directly to the population to obtain transit users.

(B) This incorrect answer results from reversing the auto density to 0.80 households per auto instead of 0.80 autos per household.

(C) This incorrect answer results from applying the transit percentage to the population density instead of to the total trips.

SOLUTION 38

A magnetic compass can show the heading under no power and poor visibility conditions. A universal default numbering system is a fail-safe feature of aviation, so that pilots can land planes with the greatest amount of safety. Without a unified numbering system, confusion would result, especially among international flights.

The answer is (D).

Why Other Options Are Wrong

(A) This answer is incorrect. Using bearings requires additional letters, N-S-E-W, and the result would be repeated numbers. Clarity of radio communications would not be good with different pronunciations of letters in different languages.

(B) This answer is incorrect. Assigning random numbers by various designers would lead to massive confusion between airports. Leaving the choice to designers could lead to the adoption of many systems. That fact, and the problem of pride of authorship, would interfere with the adoption of a universal system.

(C) This answer is incorrect. When taking off, the pilot already knows what direction the plane is heading. The need to find the correct landing heading under flight conditions is vastly more critical. The greater necessity, then, would favor landing headings.

SOLUTION 39

Effective width is determined by HCM Eq. 18-2.

$$V_p = \frac{V_{15}}{(15 \text{ min})W_E}$$

Determine the number of persons leaving in the first hour.

$$\begin{aligned}\text{persons leaving in first hour} &= (\text{person capacity}) \\ &\quad \times (\% \text{ leaving in first hour}) \\ &= (50{,}000 \text{ persons})(0.95) \\ &= 47{,}500 \text{ persons}\end{aligned}$$

Determine the transit volume.

$$\begin{aligned}\text{transit volume} &= (\text{persons leaving in first hour}) \\ &\quad \times (\% \text{ transit riders}) \\ &= (47{,}5000 \text{ persons})(0.35) \\ &= 16{,}625 \text{ transit riders}\end{aligned}$$

Define the peak hour factor (PHF), where V_{15} is the flow rate in persons per hour during the peak 15 min period.

$$\text{PHF} = \frac{V}{4V_{15}}$$

Determine the peak 15 min volume.

$$V_{15} = \frac{V}{4(\text{PHF})} = \frac{16{,}625 \text{ transit riders}}{(4)(0.75)}$$
$$= 5542 \text{ transit riders}$$

Find the flow rate from HCM Exh. 18-4. The minimum width for LOS D utilizes the maximum flow rate of 11 peds-ft/min.

Determine the effective width required using HCM Eq. 18-2.

$$W_E = \frac{V_{15}}{15V_p} = \frac{5542 \text{ peds}}{(15 \text{ min})\left(11 \ \frac{\text{peds-ft}}{\text{min}}\right)}$$
$$= 33.6 \text{ ft} \quad (34 \text{ ft})$$

The answer is (B).

Why Other Options Are Wrong

(A) This incorrect answer uses the flow rate for LOS D, unadjusted for platoons, as described in HCM prior to the 2000 edition.

(C) This incorrect answer results from a misapplication of the PHF.

(D) This incorrect answer results from determining the width needed for the entire stadium exit in the peak hour.

SOLUTION 40

The number of spaces can be determined by dividing the total lot area by the area required for one space, which includes a portion of the adjoining aisle needed for that space.

$$\text{no. of spaces} = \left(\frac{\text{total lot area}}{\text{area needed per space}}\right)\left(\text{proportion of area usable}\right)$$

Determine the area required for two spaces and the adjoining center aisle.

$$\begin{aligned}A_2 &= S_w(2S_l + I_w) \\ &= (8.5 \text{ ft})(18 \text{ ft} + 24 \text{ ft} + 18 \text{ ft}) \\ &= 510 \text{ ft}^2\end{aligned}$$

The area required for one space is

$$A_1 = \frac{A_2}{2} = \frac{510 \text{ ft}^2}{2}$$
$$= 255 \text{ ft}^2$$

The net available area for parking after deducting for maneuvering and driveway space is

$$A_{\text{net}} = A\begin{pmatrix}\text{proportion of}\\ \text{area usable}\end{pmatrix} = \left(43{,}560 \, \frac{\text{ft}^2}{\text{ac}}\right)(0.95)$$
$$= 41{,}382 \text{ ft}^2/\text{ac}$$

Determine the number of parking spaces available per acre.

$$\text{no. of spaces} = \frac{41{,}382 \, \dfrac{\text{ft}^2}{\text{ac}}}{255 \, \dfrac{\text{ft}^2}{\text{space}}}$$
$$= 162 \text{ spaces/ac} \quad (160 \text{ spaces/ac})$$

The answer is (C).

Why Other Options Are Wrong

(A) This answer incorrectly assumes that the initial area determination is for one space, not two.

(B) This answer incorrectly attaches to one space the full width of the adjacent aisle.

(D) This incorrect answer fails to reduce the available parking area by the maneuvering and aisle area.

SOLUTION 41

The total area required, A_{total}, is the sum of the areas required for each space required by type, A_i, multiplied by the number of spaces required by type, N_i. The sum is adjusted by a factor of utilization, f_u, which accounts for space needed for maneuvering and driveway areas.

$$A_{\text{total}} = \frac{\sum N_i A_i}{f_u}$$

Determine the area required for one normal space, A_n.

$$A_n = S_w\left(S_l + \frac{I_w}{2}\right) = (9 \text{ ft})\left(18 \text{ ft} + \frac{22 \text{ ft}}{2}\right)$$
$$= 261 \text{ ft}^2/\text{space}$$

Determine the total handicapped spaces required, H.

$$H = \begin{pmatrix}\text{percentage of handicapped}\\ \text{spaces required}\end{pmatrix}$$
$$\times \begin{pmatrix}\text{no. of}\\ \text{spaces provided}\end{pmatrix}$$
$$= (0.02)(550 \text{ space})$$
$$= 11 \text{ spaces}$$

Two adjoining handicapped auto spaces may share the same side aisle.

Determine the area for two adjacent handicapped auto spaces, $A_{2,\text{han}}$.

$$A_{2,\text{han}} = (S_{w,\text{han}} + I_{w,\text{han}} + S_{w,\text{han}})\left(S_l + \frac{I_w}{2}\right)$$
$$= (8 \text{ ft} + 5 \text{ ft} + 8 \text{ ft})\left(18 \text{ ft} + \frac{22 \text{ ft}}{2}\right)$$
$$= 609 \text{ ft}^2/2 \text{ spaces}$$

Determine the area for one handicapped space with an access aisle, A_{han}.

$$A_{\text{han}} = (S_{w,\text{han}} + I_{w,\text{han}})\left(S_l + \frac{I_w}{2}\right)$$
$$= (8 \text{ ft} + 5 \text{ ft})\left(18 \text{ ft} + \frac{22 \text{ ft}}{2}\right)$$
$$= 377 \text{ ft}^2/1 \text{ space}$$

Two handicapped van spaces, V_2, are required. Therefore, they can share the same access aisle.

Determine the area for the two handicapped van spaces, $A_{2,\text{van}}$.

$$A_{2,\text{van}} = (S_{w,\text{van}} + I_{w,\text{van}} + S_{w,\text{van}})\left(S_l + \frac{I_w}{2}\right)$$
$$= (8 \text{ ft} + 8 \text{ ft} + 8 \text{ ft})\left(18 \text{ ft} + \frac{22 \text{ ft}}{2}\right)$$
$$= 696 \text{ ft}^2/2 \text{ spaces}$$

Compute the total lot area required.

$$A_{\text{total}} = \frac{\begin{array}{c}NA_n + H_2 A_{2,\text{han}}\\ + HA_{\text{han}} + V_2 A_{2,\text{van}}\end{array}}{f_u}$$

$$= \frac{\begin{array}{c}(550 \text{ spaces} - 11 \text{ spaces})\left(261 \, \dfrac{\text{ft}^2}{\text{space}}\right)\\ +\begin{pmatrix}11 \text{ handicapped spaces} - 2 \text{ van spaces}\\ -1 \text{ handicapped space}\end{pmatrix}\\ \times \left(\dfrac{609 \text{ ft}^2}{2 \text{ spaces}}\right) + (1 \text{ handicapped space})\\ \times \left(\dfrac{377 \text{ ft}^2}{\text{space}}\right) + (2 \text{ van spaces})\left(\dfrac{696 \text{ ft}^2}{2 \text{ spaces}}\right)\end{array}}{0.97}$$

$$= 148{,}647 \text{ ft}^2 \quad (149{,}000 \text{ ft}^2)$$

The answer is (B).

Why Other Options Are Wrong

(A) This incorrect answer is the total parking space and aisle area required without including the maneuvering and driveway area.

(C) This incorrect answer results from adding the number of handicapped spaces to the total required instead of including the handicapped spaces in the total.

(D) This incorrect answer results from including the entire aisle width adjacent to a space as the space requirement instead of using one-half of the aisle width.

SOLUTION 42

The number of spaces can be found by dividing the total lot area by the area needed for each space plus the portion of the aisle needed for that space, then adjusting for driveways and maneuvering areas.

Determine the area required for a module of two spaces and the adjoining shared aisle.

$$A_{2,\text{sp}} = S_{wp}(2S_d + I_w)$$
$$= (9.8 \text{ ft})((2)(18 \text{ ft}) + 18 \text{ ft})$$
$$= 529.2 \text{ ft}^2/\text{space}$$

The area required for one space is

$$A_{\text{sp}} = \frac{529.2 \text{ ft}^2}{2 \text{ spaces}}$$
$$= 264.6 \text{ ft}^2/\text{space}$$

Determine the total number of spaces possible.

$$\text{no. of spaces possible} = \frac{\left(43{,}560 \frac{\text{ft}^2}{\text{ac}}\right)(1.5 \text{ ac})}{264.6 \frac{\text{ft}^2}{\text{space}}}$$
$$= 247 \text{ spaces}$$

Adjust for driveways and maneuvering space.

$$\begin{pmatrix}\text{adjusted no. of} \\ \text{spaces possible}\end{pmatrix} = \begin{pmatrix}\text{no. of} \\ \text{spaces possible}\end{pmatrix}$$
$$\times \begin{pmatrix}\text{proportion of} \\ \text{area usable}\end{pmatrix}$$
$$= (247 \text{ spaces})(0.90)$$
$$= 222 \text{ spaces}$$

The answer is (B).

Why Other Options Are Wrong

(A) This incorrect answer results from using the aisle module unit area for one space instead of for two spaces.

(C) This incorrect answer results from failing to adjust for maneuvering and driveway area.

(D) This incorrect answer results from using the right-angle space width, S_w, which is the wrong dimension, to determine the module area required per space.

SOLUTION 43

The total number of parking spaces required is the sum of spaces required for each assigned building use.

$$N_{\text{total}} = N_o + N_r + N_{w/s}$$

Determine the space assigned to mechanical uses, for which no parking spaces are required.

$$\text{mechanical space assigned} = (80{,}000 \text{ ft}^2)(0.10)$$
$$= 8000 \text{ ft}^2$$

Determine the warehouse and storage space.

$$\text{space assignment} = 80{,}000 \text{ ft}^2 - 45{,}000 \text{ ft}^2$$
$$- 12{,}000 \text{ ft}^2 - 8000 \text{ ft}^2$$
$$= 15{,}000 \text{ ft}^2$$

Determine the required parking spaces for each use.

For offices,

$$N_o = \frac{45{,}000 \text{ ft}^2}{300 \frac{\text{ft}^2}{\text{space}}}$$
$$= 150 \text{ spaces}$$

For retail use,

$$N_r = \frac{12{,}000 \text{ ft}^2}{150 \frac{\text{ft}^2}{\text{space}}}$$
$$= 80 \text{ spaces}$$

For warehouse and storage use,

$$N_{w/s} = \frac{15{,}000 \text{ ft}^2}{500 \frac{\text{ft}^2}{\text{space}}}$$
$$= 30 \text{ spaces}$$
$$N_{\text{total}} = N_o + N_r + N_{w/s}$$
$$= 150 \text{ spaces} + 80 \text{ spaces} + 30 \text{ spaces}$$
$$= 260 \text{ spaces}$$

The answer is (C).

Why Other Options Are Wrong

(A) This incorrect answer is for the office spaces only. No specific information was given in the problem to allow consideration of office space and retail space having separate hours. Therefore, overlapping hours must be considered, requiring full accommodation of both.

(B) This incorrect answer does not include parking spaces needed for the warehouse and storage occupancy.

(D) This incorrect answer assigns parking spaces to the mechanical space as if it were warehouse and storage space. Mechanical space is not required to have parking space allocation.

SOLUTION 44

The number of lanes required, n, is determined by the maximum flow rate divided by the service rate, μ, of each lane.

$$n = \frac{V_p}{\mu}$$

The maximum flow rate, V_p, is determined by the peak-hour arrival of cars, V, divided by the PHF.

$$V_p = \frac{V}{\text{PHF}}$$

Combine the two relationships to determine n.

$$n = \frac{V}{(\text{PHF})\mu}$$

Determine the number of cars arriving in the design hour.

$$V = (\text{vehicles arriving in design hour})(\text{lot capacity})$$
$$= \left(\frac{0.72}{\text{hr}}\right)(700 \text{ veh})$$
$$= 504 \text{ vph}$$

Compare the required n for each PHF value.

$$n_1 = \frac{504 \ \frac{\text{veh}}{\text{hr}}}{(0.70)\left(270 \ \frac{\text{veh}}{\text{hr}}\right)} = 2.7$$

$$n_2 = \frac{504 \ \frac{\text{veh}}{\text{hr}}}{(0.80)\left(270 \ \frac{\text{veh}}{\text{hr}}\right)} = 2.3$$

$$n_3 = \frac{504 \ \frac{\text{veh}}{\text{hr}}}{(0.87)\left(270 \ \frac{\text{veh}}{\text{hr}}\right)} = 2.1$$

$$n_4 = \frac{504 \ \frac{\text{veh}}{\text{hr}}}{(0.94)\left(270 \ \frac{\text{veh}}{\text{hr}}\right)} = 2.0$$

$$n_5 = \frac{504 \ \frac{\text{veh}}{\text{hr}}}{(0.97)\left(270 \ \frac{\text{veh}}{\text{hr}}\right)} = 1.9$$

Summarizing,

no. of lanes	PHF	$(\text{PHF})\mu$ (vph)	n_{required} (lanes)
1	0.70	189	2.7
2	0.80	216	2.3
3	0.87	235	2.1
4	0.94	254	2.0
5 or more	0.97	265	1.9

The equation balances at three lanes.

The answer is (B).

Why Other Options Are Wrong

(A) This incorrect answer does not adjust the service rate by the peak hour factor.

(C) This answer incorrectly uses the total lot capacity as the incoming volume during the design hour.

(D) This answer results from incorrectly assuming that the peak hour factor is the actual flow proportion during the peak 15 min period instead of the flow *rate* during the peak 15 min period.

SOLUTION 45

Determine the degree of saturation for each phase using

$$x_i = \frac{V_i}{C_i}$$

V is the volume of the ith approach and C is the capacity of the ith approach.

For phase A,

$$x_{\text{A}} = \frac{V_{\text{A}}}{C_{\text{A}}} = \frac{70 \ \frac{\text{veh}}{\text{hr}}}{350 \ \frac{\text{veh}}{\text{hr}}}$$
$$= 0.200$$

For phase B,

$$x_{\text{B}} = \frac{V_{\text{B}}}{C_{\text{B}}} = \frac{400 \ \frac{\text{veh}}{\text{hr}}}{1350 \ \frac{\text{veh}}{\text{hr}}}$$
$$= 0.296$$

For phase C,

$$x_C = \frac{V_C}{C_C} = \frac{550 \ \frac{\text{veh}}{\text{hr}}}{1960 \ \frac{\text{veh}}{\text{hr}}}$$
$$= 0.281$$

The critical sum for the cycle is the sum of the flow ratios of the critical movements for all phases.

$$\text{CS} = \sum x_i$$
$$= 0.200 + 0.296 + 0.281$$
$$= 0.777$$

Apply Webster's equation, which divides the factored sum of the phase lost times by the unused cycle saturation.

$$C_O = \frac{(1.5)\,L + 5 \text{ sec}}{1 - \text{CS}}$$
$$= \frac{(1.5)\,((3 \text{ sec})\,(3)) + 5 \text{ sec}}{1 - 0.777}$$
$$= 83.0 \text{ sec}$$

C_O is the optimal cycle length and L is the lost time per cycle (the lost time per phase in seconds times the number of phases).

The answer is (D).

Why Other Options Are Wrong

(A) In this incorrect answer, failure to subtract the sum of the critical flow ratios from 1 in the equation denominator results in too short of a cycle length.

(B) In this incorrect answer, using the ratio of the sum of the critical flows instead of the sum of the critical flow ratios results in too short of a cycle length.

(C) In this incorrect answer, not including the lost time for *each* phase, but only including the 3 sec lost time for the entire cycle, results in too short of a cycle length.

SOLUTION 46

Saturation flow rate is determined from HCM Eq. 16-4.

$$s = s_o N f_w f_{\text{HV}} f_g f_p f_{\text{bb}} f_a f_{\text{LU}} f_{\text{LT}} f_{\text{RT}} f_{L\text{pb}} f_{R\text{pb}}$$
$$s_o = 1900 \text{ pcphpl}$$
$$N = 1$$
$$f_w = 1 + \frac{w - 12}{30}$$
$$= 1 + \frac{11 - 12}{30}$$
$$= 0.97$$

$$f_{\text{HV}} = \frac{100}{100 + \%\text{HV}(E_T - 1)}$$
$$= \frac{100}{100 + (0)(2 - 1)}$$
$$= 1.00$$
$$f_g = 1 - \frac{\%G}{200}$$
$$= 1 - \frac{4.0}{200}$$
$$= 0.98$$
$$f_p = \frac{N - 0.1 - \frac{18 N_m}{3600}}{N}$$
$$= \frac{1 - 0.1 - \frac{(18)(20)}{3600}}{1}$$
$$= 0.80$$
$$f_{\text{bb}} = \frac{N - \frac{14.4 N_B}{3600}}{N}$$
$$= \frac{1 - \frac{(14.4)(0)}{3600}}{1}$$
$$= 1.00$$
$$f_a = 0.90 \text{ [for CBD]}$$
$$f_{\text{LU}} = \frac{V_g}{V_{g1} N}$$
$$= \frac{731 + 139}{(731 + 139)(1)}$$
$$= 1.00$$
$$f_{\text{LT}} = \frac{1}{1.0 + 0.05 P_{\text{LT}}}$$
$$= \frac{1}{1.0 + (0.05)(0)}$$
$$= 1.00$$
$$f_{\text{RT}} = 1.0 - 0.135 P_{\text{RT}}$$
$$= 1.0 - (0.135)\left(\frac{139}{870}\right)$$
$$= 0.98$$
$$f_{L\text{pb}} = 1.00$$
$$f_{R\text{pb}} = 1.00$$

Determine the saturation flow rate.

$$s = (1900 \text{ pcphpl})(1)(0.97)(1.0)(0.98)(0.80)(1.0)$$
$$\times (0.90)(1.0)(1.0)(0.98)(1.0)(1.0)$$
$$= 1274 \text{ vph} \quad (1270 \text{ vph})$$

The answer is (B).

Why Other Options Are Wrong

(A) This incorrect answer is the sum of the intersection movements for approach E, or the approach volume.

(C) This incorrect answer is the result of neglecting to consider the approach grade.

(D) This incorrect answer is the result of considering the area type factor as 1.0 for CBD instead of 0.90.

SOLUTION 47

The MUTCD is very clear that simply meeting the minimum warrants for a traffic signal is not sufficient to satisfy the requirements for installation. The *Traffic Engineering Handbook* further states that "...the requirements for a signal should be thoroughly analyzed with a decision to install based on a demonstrated traffic need."

Regardless of the minimum requirements that are met by traffic conditions, an engineering study is still needed to indicate that the installation will improve overall safety of the intersection.

The answer is (A).

Why Other Options Are Wrong

(B) This answer is incorrect. Minimum pedestrian volume is to be 100 persons for each of any 4 hr period. Therefore, 100 persons for a 2 hr period is insufficient justification on its own.

(C) This answer is incorrect. Accident experience is to be considered if the accidents were of the nature that a traffic signal would be designed to avoid. Drivers not stopping at a stop sign would not necessarily stop for a traffic signal either. More information needs to be known about the nature of the accidents.

(D) This answer is incorrect. A traffic jam in the evening caused by employee discharge from work can be reduced by alternate means, such as varying the quitting times of employees. Alternate means would have to be employed and observed before declaring that a signal is justified.

SOLUTION 48

To avoid a possible moving conflict or collision, all signal faces should flash red, meaning all approach traffic is to stop. Exceptions are to show a flashing yellow for the major movement and for any nonconflicting turn movements that have adequate sight distance to proceed without requiring a full stop. It is unsafe to have more than one conflicting approach with yellow indication.

The answer is (B).

Why Other Options Are Wrong

(A) This answer is incorrect. Allowing one approach (usually the major street) to proceed without stopping is permissible.

(C) This answer is incorrect. Every movement that normally would be controlled with a R-Y-G signal would also be controlled with the red or yellow flashing signal.

(D) This answer is incorrect. This statement is true because all signal faces that are flashed on an approach are to flash the same color, except for separate signal faces for protected or protected/permissive left-turn movements, which are to flash red when the through movement signal faces flash yellow (MUTCD section 40.11).

TRANSPORTATION PLANNING

SOLUTION 49

Determine the total vehicle miles traveled in one direction.

$$\text{VM}_{\text{total}} = s\left(P_c P_{\text{total}}\left(\frac{1}{O_c}\right) + P_{\text{sov}} P_{\text{total}}\left(\frac{1}{O_{\text{sov}}}\right)\right)$$

$$= (8 \text{ mi})\left((0.25)(7000 \text{ pers})\left(\frac{1}{2.1 \frac{\text{pers}}{\text{veh}}}\right)\right.$$

$$\left. + (0.75)(7000 \text{ pers})\left(\frac{1}{1 \frac{\text{pers}}{\text{veh}}}\right)\right)$$

$$= 48{,}667 \text{ veh-mi}$$

Determine the fuel consumption rate per vehicle.

$$F = 0.0362 \ \frac{\text{gal}}{\text{veh-mi}} + \left(\frac{0.746 \ \frac{\text{gal}}{\text{veh-hr}}}{20 \ \frac{\text{mi}}{\text{hr}}}\right)$$

$$= 0.0735 \text{ gal/veh-mi}$$

Determine the gallons of fuel consumed.

$$\text{total fuel consumption} = (\text{total vehicle miles})$$
$$\times (\text{consumption rate})$$
$$= (48{,}667 \text{ veh-mi})$$
$$\times \left(0.0735 \ \frac{\text{gal}}{\text{veh-mi}}\right)$$
$$= 3577 \text{ gal}$$

Determine the Btu's consumed for a one-way trip.

$$\text{Btu consumption} = (\text{total gallons consumed})$$
$$\times \left(\frac{\text{Btu content}}{\text{gallon}}\right)$$
$$= (3577 \text{ gal})\left(125{,}000 \ \frac{\text{Btu}}{\text{gal}}\right)$$
$$= 4.47 \times 10^8 \text{ Btu}$$

This result is for a one-way trip. Each day involves two similar trips, one trip to go to work and one trip to return home. The daily consumption is

$$\begin{aligned}\text{total daily} \atop \text{consumption} &= \left(\frac{\text{number of one-way trips}}{\text{round trip}}\right) \\ &\quad \times \left(\frac{\text{Btu consumption}}{\text{one-way trip}}\right) \\ &= (2)(4.47 \times 10^8 \text{ Btu}) \\ &= 8.94 \times 10^8 \text{ Btu} \quad (8.9 \times 10^8 \text{ Btu})\end{aligned}$$

The answer is (C).

Why Other Options Are Wrong

(A) This incorrect answer is the number of gallons consumed per day.

(B) This incorrect answer is the number of Btu's consumed in a one-way trip.

(D) This incorrect answer is the number of Btu's consumed if all 7000 commuters arrived in a single-occupant vehicle.

SOLUTION 50

Improvement to public safety in urban areas involves providing an adequate physical environment so that individuals do not feel insecure or uncertain about what is happening nearby. All of the listed items can improve the feeling of personal comfort or reduce the level of discomfort with surroundings. The traveling public is surrounded by noise and vibration in an urban setting at a fairly constant level. Conceivably, the noise level in a transit terminal, option (B), would not be much different from that on the street. The other three options could be perceived as making a greater difference within the transit terminal than in the surrounding neighborhood, thereby improving the safety of the terminal itself.

The answer is (B).

Why Other Options Are Wrong

(A) This answer is incorrect. Improving personal security is an inherent goal of improving public safety.

(C) This answer is incorrect. Adequate lighting levels significantly improve the travelers' confidence in their surroundings and lead to an improved feeling of personal safety.

(D) This answer is incorrect. Reliability in transit service reduces anxiety about unknown arrival times. This reduces the need for longer waiting times in the terminal, especially during off-peak travel times.

SOLUTION 51

Determine the number of person trips per household.

$$\begin{aligned}T &= 0.78 + 1.3P + 2.3A \\ &= 0.78 + (1.3)\left(3.5 \frac{\text{persons}}{\text{household}}\right) \\ &\quad + (2.3)\left(2.2 \frac{\text{autos}}{\text{household}}\right) \\ &= 10.39 \text{ trips/household-day}\end{aligned}$$

Determine the number of person trips in the entire zone.

$$\begin{aligned}N &= T(\text{no. of households}) \\ &= \left(10.39 \frac{\text{trips}}{\text{household-day}}\right)(600 \text{ households}) \\ &= 6234 \text{ trips/day}\end{aligned}$$

Determine the number of auto trips per day.

$$\begin{aligned}T_A &= (\text{proportion of auto trips})\left(\frac{\text{total trips}}{\text{day}}\right) \\ &= (0.94)\left(6234 \frac{\text{trips}}{\text{day}}\right) \\ &= 5860 \text{ trips/day} \quad (5900 \text{ trips/day})\end{aligned}$$

The answer is (B).

Why Other Options Are Wrong

(A) This incorrect answer is the number of trips per household.

(C) This incorrect answer is the total number of trips per day for the zone, including transit and other.

(D) This incorrect answer results from misuse of the modal split ratio.

CONSTRUCTION

SOLUTION 52

The average end area and length must be determined for each section and summed to determine the earthwork volumes. Determine the average end area.

For cut,

$$\text{sta 1 to 2} = \frac{0 \text{ ft}^2 + 250 \text{ ft}^2}{2} = 125 \text{ ft}^2$$

$$\text{sta 2 to 3} = \frac{250 \text{ ft}^2 + 100 \text{ ft}^2}{2} = 175 \text{ ft}^2$$

For fill,

$$\text{sta 1 to 2} = \frac{50 \text{ ft}^2 + 70 \text{ ft}^2}{2} = 60 \text{ ft}^2$$

$$\text{sta 2 to 3} = \frac{70 \text{ ft}^2 + 0 \text{ ft}^2}{2} = 35 \text{ ft}^2$$

Determine earthwork volumes.

For cut,
$$V = (\text{sta 1 to 2})(100\text{ ft}) + (\text{sta 2 to 3})(100\text{ ft})$$
$$= (125\text{ ft}^2)(100\text{ ft}) + (175\text{ ft}^2)(100\text{ ft})$$
$$= 30{,}000\text{ ft}^3$$

For fill,
$$V = (\text{sta 1 to 2})(100\text{ ft}) + (\text{sta 2 to 3})(100\text{ ft})$$
$$= (60\text{ ft}^2)(100\text{ ft}) + (35\text{ ft}^2)(100\text{ ft})$$
$$= 9500\text{ ft}^3$$

There is more cut than fill; therefore, this is a waste job. Determine the amount of waste.

$$\text{amount of waste} = \text{cut } V - \text{fill } V$$
$$= (30{,}000\text{ ft}^3 - 9500\text{ ft}^3)\left(\frac{1\text{ yd}^3}{27\text{ ft}^3}\right)$$
$$= 759\text{ yd}^3 \quad (760\text{ yd}^3)$$

The answer is (C).

Why Other Options Are Wrong

(A) This incorrect answer is the amount of fill.

(B) This incorrect answer is the amount of difference between cut and fill; however, borrow means to bring material onto the job rather than haul it away.

(D) This incorrect answer is the total amount of cut, not reduced by the fill volume needed.

SOLUTION 53

The new side slopes will remove a triangular wedge of material beyond the originally proposed slope.

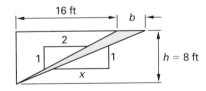

Determine the additional cut volume required per foot of roadway.

$$V = \frac{\text{volume of borrow}}{\text{length of roadway}} = \frac{(2000\text{ yd}^3)\left(27\,\frac{\text{ft}^3}{\text{yd}^3}\right)}{3500\text{ ft}}$$
$$= 15.429\text{ ft}^3/\text{ft}$$

The additional cut will be made equally on both sides.

$$\frac{V}{2\text{ sides}} = \frac{15.429\,\frac{\text{ft}^3}{\text{ft}}}{2\text{ sides}}$$
$$= 7.715\text{ ft}^3/\text{side}$$

The volume per side, 7.715 ft², is the area of the additional triangular wedge, 1 ft in thickness.

Determine the top width, or base, of the triangle using the formula
$$A = \tfrac{1}{2}bh$$

Rearrange to find the base.
$$b = \frac{2A}{h} = \frac{(2)(7.715\text{ ft}^2)}{8\text{ ft}}$$
$$= 1.93\text{ ft}$$

Find the new slope, x.
$$x = \text{horizontal measure:vertical measure}$$
$$= (16\text{ ft} + 1.93\text{ ft}):8\text{ ft}$$
$$= 2.24{:}1$$

The answer is (B).

Why Other Options Are Wrong

(A) This incorrect answer results from reversing the side-slope designation of the original cross section. It should be the tangent of the slope angle from vertical, not the tangent of the angle from horizontal.

(C) This incorrect answer results from assigning all of the required overexcavation to one side instead of equally to both sides.

(D) This incorrect answer results from improperly assigning the base dimension to the diagonal altitude of the overexcavation triangular wedge.

SOLUTION 54

The limit of economical haul is the sum of the economical length of overhaul plus the length of freehaul.

$$\text{leh} = \text{loh} + \text{lfh}$$

The limit of freehaul is given as 800 ft; therefore, the unknown is the limit of overhaul.

The economical limit of overhaul is met when the cost of overhaul is equal to the cost of excavation.

$$\text{loh} = \frac{c}{h} = \frac{\frac{\$2.70}{\text{yd}^3}}{\frac{\$0.90}{\text{yd}^3\text{-sta}}} = 3\text{ sta} \quad (300\text{ ft})$$

$$\text{leh} = 300\text{ ft} + 800\text{ ft} = 1100\text{ ft}$$

The answer is (D).

Why Other Options Are Wrong

(A) Should the excavation cost and overhaul cost be inverted in the calculation of the economical length of overhaul and confused with the limit of economical haul, the resulting incorrect answer will be too small.

(B) This incorrect answer correctly states that the economical limit of overhaul is 300 ft, but it does not account for the fact that the total economical limit of haul also includes the freehaul.

(C) This incorrect answer fails to regard the total economical limit of haul as the freehaul length of 800 ft plus the economical limit of overhaul of 300 ft.

SOLUTION 55

The EAL for each group is determined by

$$\text{EAL} = (\text{veh})(\text{TF})(\text{GF})$$

Determine the growth-rate factor.

$$\text{GF} = \frac{(1+i)^n - 1}{i} = \frac{(1+0.03)^{15} - 1}{0.03}$$
$$= 18.6$$

Determine the design EAL total.

vehicle type	number of vehicles (per year)	truck factor	15 yr growth	EAL
single units				
2-axle, 4-tire	80,000 × 0.002	× 18.6	=	2976
2-axle, 6-tire	9000 × 0.24	× 18.6	=	40,176
3-axle or more	2000 × 1.02	× 18.6	=	37,944
all single units	91,000			
tractor semitrailer				
5-axle	3000 × 0.97	× 18.6	=	54,126
all vehicles	94,000			
design EAL total				135,222 (135,000)

The answer is (C).

Why Other Options Are Wrong

(A) This incorrect EAL total results from failing to apply the growth factor to the solution.

(B) This incorrect EAL total is obtained when a simple annual increase is used instead of a compound annual increase.

(D) This incorrect EAL total results from failing to apply truck factors to the solution.

SOLUTION 56

Table 11.7 of *The Asphalt Handbook* indicates a 2 in minimum thickness over Type II and Type III base courses for light traffic.

The answer is (B).

(A) This answer is incorrect. A 1 in minimum thickness is only recommended over Type I base courses (i.e., over very sound material such as concrete) and only for light traffic conditions.

(C) This answer is incorrect. *The Asphalt Handbook* has no recommendation for a $2^{1}/_{2}$ in surface base course.

(D) This answer is incorrect. A 3 in minimum thickness is recommended over Type II or Type III base course material for medium (not light) traffic conditions.

SOLUTION 57

The volume per linear foot of the windrow is the volume of the in-place aggregate per linear foot divided by the loose bulking factor. The loose bulking factor compares the loose density of the uncompacted windrow with the in-place compacted density.

Determine the in-place volume of aggregate per linear foot of roadway.

$$V = hwL = (4 \text{ in})\left(\frac{1 \text{ ft}}{12 \text{ in}}\right)(11 \text{ ft})(1 \text{ ft})$$
$$= 3.67 \text{ ft}^3/\text{ft}$$

Determine the bulk volume of the windrow needed per linear foot of roadway using a bulking factor, BF, of 0.68.

$$V_b = V\left(\frac{1}{\text{BF}}\right) = \left(3.67 \, \frac{\text{ft}^3}{\text{ft}}\right)\left(\frac{1}{0.68}\right)$$
$$= 5.40 \text{ ft}^3/\text{ft}$$

Determine the required volume of the windrow per linear foot of roadway.

$$V_b = \left(\frac{A+B}{2}\right)HL$$
$$= \left(\frac{A+B}{2}\right)(18 \text{ in})\left(\frac{1 \text{ ft}}{12 \text{ in}}\right)(1 \text{ ft})$$
$$= 5.4 \text{ ft}^3$$

Substitute $2A$ for B.

$$5.4 \text{ ft}^3 = \left(\frac{A+2A}{2}\right)(1.5 \text{ ft}^2)$$
$$A = 2.4 \text{ ft}$$
$$B = 2A = (2)(2.4 \text{ ft})$$
$$= 4.8 \text{ ft}$$

The answer is (D).

Why Other Options Are Wrong

(A) This incorrect answer results from multiplying (rather than dividing) the in-place density by the bulking factor.

(B) This incorrect answer is the width of the windrow top. The bottom is twice the width of the top.

(C) This incorrect answer is the windrow volume using the in-place density of aggregate in the windrow instead of the bulk volume.

SOLUTION 58

Referring to *Superpave Mix Design*, the description of Superpave mix design involves laboratory compaction and performance testing. Asphalt binders are performance graded by a term such as PG 58-22. The first number, 58, is called the high temperature grade, or the average seven-day maximum design temperature in degrees celsius. The second number is the low temperature grade, or the minimum pavement design temperature in degrees celsius. Other binder design procedures consider aging specifications. Load-carrying designs use prospective equivalent single-axle loads (ESALs).

The answer is (B).

Why Other Options Are Wrong

(A) This incorrect answer is a description of the Marshal design procedure.

(C) This incorrect answer is a description of the Hveem pavement design method.

(D) There is no such test as described. Asphalt pavement design load capacity is related to expected traffic conditions and expected ESALs, not to the maximum load at a given temperature.

SOLUTION 59

Asphalt bonding to aggregate is more of a function of aggregate porosity and surface adhesion than of alkalinity. *The Asphalt Handbook* does not specify a test for alkalinity on aggregate.

The answer is (B).

Why Other Options Are Wrong

(A) The depth of subsurface utilities may limit total pavement depth, and utilities that are in poor condition may need to be dug up frequently for repairs. Shallow pavement depth and frequent cutting for utility trenches have a significant effect on the strength and longevity of the pavement surface. This statement is true.

(C) Adjoining properties may have high axle loadings and require sharp turns by heavy vehicles, which would have a significant effect on the durability of the pavement. Pavement depth may have to be increased to account for the turning movements. Access to adjoining driveways may be needed during construction, requiring the contractor to place asphalt in shallow lifts to allow quick compaction and early traffic loads. This statement is true.

(D) The most economical pavement designs are generally ones that can be applied using existing available equipment and that are within the skills of available contractors. The use of designs requiring extraordinary skill or equipment needs to take into account the extra effort requirement for success of the project. This statement is true.

SOLUTION 60

Determine the percentage passing and cumulative percentage retained by continuing the table.

sieve size	percentage of sample retained, by weight (%)	cumulative percentage passing (%)	cumulative percentage retained (%)
3/8 in	1	99	1
no. 4	3	96	4
no. 8	11	85	15
no. 16	20	65	35
no. 30	24	41	59
no. 50	22	19	81
no. 100	17	2	98
pan	2	0	–
total	100	–	293

Determine the sum of the cumulative percentage retained on each sieve, not counting the amount in the pan (2% is negligible).

$$\sum \text{cumulative \% retained} = 1\% + 4\% + 15\% + 35\% \\ + 59\% + 81\% + 98\% \\ = 293\%$$

Divide by 100% to obtain the fineness modulus, FM.

$$\text{FM} = \frac{\sum \text{cumulative \% retained}}{100\%} \\ = \frac{293\%}{100\%} \\ = 2.93 \quad (2.9)$$

The answer is (B).

Why Other Options Are Wrong

(A) This incorrect answer is the result of adding the total number of samples retained on each sieve and then dividing by 100.

(C) This incorrect answer is the result of adding the cumulative percentage retained, including the pan value.

(D) This incorrect value is the result of adding the cumulative amount passing through the sieves.

SOLUTION 61

Heating the aggregate in a drum mixer above 100°C drives out water, both surface and absorbed. For hot-mix asphalt, the aggregate is dried using superheated air before the asphalt liquid is introduced. Simply air drying aggregate at ambient temperature will not remove all of the moisture. Therefore, option (D) is false.

The answer is (D).

Why Other Options Are Wrong

(A) SSD is a common design parameter of specifying concrete mix proportions. This statement is true.

(B) During hot-mix aggregate preparation, the aggregate is heated to about 300°F (150°C before liquid hot asphalt is introduced), which drives out any moisture. Therefore, the moisture content of the raw aggregate has no effect on the mix design. This statement is true.

(C) Damp or wet aggregate is defined as the condition in which the aggregate contains enough moisture to saturate, plus excess moisture on the surface of the aggregate pieces. This statement is true.

SOLUTION 62

Options (A), (C), and (D) mention an easily determined internal dimension of the finished member. The volume of air voids decreases as the large aggregate size is increased. The air-void volume is normally less than 10%, even with large aggregate as small as 3/8 in. (see *Design and Control of Concrete Mixtures* Ch. 8). Therefore, option (B) is incorrect.

The answer is (B).

Why Other Options Are Wrong

(A) This criterion is specified by ACI 318-3.3. Therefore, the statement is true.

(C) This criterion is specified by ACI 318-3.3. Therefore, the statement is true.

(D) This criterion is specified by ACI 318-3.3. Therefore, the statement is true.

SOLUTION 63

Substances deposited in the aggregate that reduce essential properties and weaken the mix are considered deleterious. These would include bits of wood, soil, organic materials, or even chemical contaminants (see *Design and Control of Concrete Mixtures* Ch. 5). Therefore, option (C) is false.

The answer is (C).

Why Other Options Are Wrong

(A) Most aggregate that passes other physical tests has a greater strength than the strength of the completed mix for normal-strength concrete (see PCA Ch. 5). Therefore, this statement is true.

(B) Generally, harder aggregates have a higher abrasion resistance. Higher abrasion resistance increases the abrasion resistance of the completed mix, which may be desirable for high-traffic locations (see PCA Ch. 5). This statement is true.

(D) The presence of clay or shale increases shrinking and swelling, which can reduce the life of the concrete under wetting and drying conditions (see PCA Ch. 5). This statement is true.

SOLUTION 64

Find the total cost of construction, starting with the joint repair.

$$\text{total number of joints} = (\text{length of project})\left(\frac{1}{\text{joint spacing}}\right)$$

$$= (10 \text{ mi})\left(5280 \frac{\text{ft}}{\text{mi}}\right)\left(\frac{1}{44 \frac{\text{ft}}{\text{joint}}}\right)$$

$$= 1200 \text{ joints}$$

$$\text{joint replacement cost} = (\text{proportion of joints replaced})$$
$$\times (\text{total number of joints})$$
$$\times (\text{cost of joint replacement})$$

$$= (0.15)(1200 \text{ joints})\left(\frac{\$1800}{\text{joint}}\right)$$

$$= \$324{,}000$$

$$\text{joint repair cost} = (\text{proportion of joints repaired})$$
$$\times (\text{total number of joints})$$
$$\times (\text{cost of joint repair})$$

$$= (0.85)(1200 \text{ joints})\left(\frac{\$450}{\text{joint}}\right)$$

$$= \$459{,}000$$

$$\frac{\text{total cost of}}{\text{joint work}} = \text{cost of joint replacement}$$
$$+ \text{cost of repair}$$
$$= \$324{,}000 + \$459{,}000$$
$$= \$783{,}000$$

overlay cost = (length of project)(width of project)

$$= (10 \text{ mi}) \left(5280 \, \frac{\text{ft}}{\text{mi}}\right) \left(\frac{24 \text{ ft}}{9 \, \frac{\text{ft}^2}{\text{yd}^2}}\right) \left(\frac{\$11}{\text{yd}^2}\right)$$

$$= \$1{,}548{,}800$$

$$\frac{\text{total construction}}{\text{cost}} = \text{total cost of joint work}$$
$$+ \text{overlay cost}$$
$$= \$783{,}000 + \$1{,}548{,}800$$
$$= \$2{,}331{,}800$$

Determine the annual capital recovery amount. The values for A and P are obtained from apropriate interest-rate tables.

$$A = P\left(\frac{A}{P}, i, n\right)$$
$$= (\$2{,}331{,}800)(0.1666)$$
$$= \$388{,}478/\text{yr} \quad (\$389{,}000/\text{yr})$$

The answer is (B).

Alternate Solution:

Using the mathematical formula,

$$A = P\left(\frac{i(1+i)^n}{(1+i)^n - 1}\right)$$
$$= (\$2{,}331{,}800)\left(\frac{(0.04)(1.04)^7}{(1.04)^7 - 1}\right)$$
$$= \$388{,}500/\text{yr} \quad (\$389{,}000/\text{yr})$$

Why Other Options Are Wrong

(A) This incorrect answer is the result of determining the straight-line depreciation of the initial construction cost, not considering interest.

(C) This incorrect answer is the result of incorrectly adding the cost of sawcutting and sealing joints to the cost of the joints that are to be replaced.

(D) This incorrect answer is the result of incorrectly determining the end-of-life value of the entire construction cost, then assigning a proportional amount to each year of life.

$$A = P\left(\frac{F}{P}, i, n\right) = (\$2{,}331{,}800)(1.3159)$$
$$= \$3{,}068{,}416$$

Determine the annual amount by direct proportion.

$$A = (\$3{,}068{,}416)\left(\frac{1}{7 \text{ yr}}\right)$$
$$= \$438{,}345/\text{yr} \quad (\$438{,}000/\text{yr})$$

SOLUTION 65

Determine the EAL for each load group.

$$(\text{LEF})_i \, (\% \text{ ADT})_i = \text{EAL}_i$$
$$(0.0002)_{3000} \, (0.67) \, (8500 \text{ veh}) = 1 \text{ EAL}$$
$$(1.0)_{18{,}000} \, (0.21) \, (8500 \text{ veh}) = 1785 \text{ EAL}$$
$$(0.027)_{14{,}000} \, (0.06) \, (8500 \text{ veh}) \, (2 \text{ axles}) = 28 \text{ EAL}$$
$$(0.180)_{22{,}000} \, (0.06) \, (8500 \text{ veh}) \, (2 \text{ axles}) = 184 \text{ EAL}$$

Determine the total daily EAL.

$$\text{EAL}_{\text{daily}} = \sum \text{EAL}$$
$$= 1 \text{ EAL} + 1785 \text{ EAL} + 28 \text{ EAL} + 184 \text{ EAL}$$
$$= 1998 \text{ EAL/day}$$

Determine the total annual EAL.

$$\text{EAL}_{\text{annual}} = (\text{EAL}_{\text{daily}})t = \left(1998 \, \frac{\text{EAL}}{\text{day}}\right)\left(365 \, \frac{\text{days}}{\text{yr}}\right)$$
$$= 729{,}270 \text{ EAL/yr}$$

Determine the design lane EAL.

$$\text{EAL}_{\text{design lane}} = 0.6(\text{EAL}_{\text{annual}})$$
$$= (0.6)\left(729{,}270 \, \frac{\text{EAL}}{\text{yr}}\right)$$
$$= 437{,}562 \text{ EAL/yr} \quad (440{,}000 \text{ EAL/yr})$$

The answer is (C).

Why Other Options Are Wrong

(A) This incorrect solution uses the daily EAL, whereas design charts require annual EAL.

(B) This incorrect solution uses a single axle for the 14,000 lbf axle and the 22,000 lbf axle instead of two separate axles.

(D) This incorrect solution uses the total annual EAL without considering directional distribution.

SOLUTION 66

According to Ch. 4 of *The Asphalt Handbook*, a properly designed asphalt pavement mix will contain enough asphalt to coat all of the aggregate surfaces, but not so much that the asphalt cement becomes the primary load-carrying ingredient. The final pavement structure must adjust for the design load condition and then rebound elastically without distortion. A mix that is too hard will fracture and not be able to rebound. Also, a mix that is too hard will not hold up under high volumes of traffic.

The answer is (A).

Why Other Options Are Wrong

(B) This answer is incorrect. Mixes with high flow values that distort easily under traffic loads are considered unstable. See *The Asphalt Handbook*, Table 4.4.

(C) This answer is incorrect. Reducing asphalt content may not provide enough material to coat the aggregate, thereby reducing shear strength and allowing increased fatigue cracking. See Ch. 4 of *The Asphalt Handbook*.

(D) This answer is incorrect. The maximum load applied when the test sample fails is considered the Marshal stability value, along with the amount of flow. There is no standard test to determine the maximum load without the sample failing.

SOLUTION 67

When the asphalt contacts a porous surface, such as a rubber tire, it will cool slightly and cling. A steel roller is not very porous, so the cooled asphalt will not stick if there is a slight film of water on the roller surface. The rubber tire is porous enough that, even with a water film, the asphalt will stick when cooled by the tire. When the tire is kept as hot as the asphalt, the asphalt remains fluid enough to prevent sticking. See Ch. 7 of *The Asphalt Handbook*.

The answer is (D).

Why Other Options Are Wrong

(A) Starting to roll the spread from the low side will cause uncompacted asphalt mat to nest against the compacted section rather than slip down the slope (see Ch. 7 of *The Asphalt Handbook*). This statement is true.

(B) The unsupported edge will ravel and collapse if the roller is operated too close to the edge. When asphalt is first rolled away from the edge, the row of uncompacted asphalt acts as a dike to hold the remaining mat in place. A final pass over the uncompacted dike will not slough off if the roller is supported on the compacted mat (see Ch. 7 of *The Asphalt Handbook*). This statement is true.

(C) Finish rolling is done to erase roller marks and other cosmetic imperfections, using a nonvibrating roller to avoid leaving undesirable marks in the surface and create a corrugated riding surface (see Ch. 7 of *The Asphalt Handbook*). This statement is true.

GEOMETRIC DESIGN

SOLUTION 68

Change the design speed from miles per hour to feet per second.

$$v_{ft/sec} = \left(70 \frac{mi}{hr}\right)\left(5280 \frac{ft}{mi}\right)\left(\frac{1 \text{ hr}}{3600 \text{ sec}}\right)$$
$$= 102.7 \text{ ft/sec}$$

The roadway is already sloped down to the right at 0.01 ft/ft, and the superelevation will increase the slope of the roadway down to the right since this is a right-hand curve. Therefore, there will be no reverse cross slope to run out on the right side. The difference between normal slope and full superelevation is

$$\text{required slope transition} = 0.08 \frac{ft}{ft} - 0.01 \frac{ft}{ft} = 0.07 \text{ ft/ft}$$

The number of seconds required to transition from normal slope to full superelevation is determined by dividing that transition slope change by the allowable slope change rate.

$$\text{min transition time} = \frac{\text{required slope transition}}{\text{slope change rate}}$$
$$= \frac{0.07 \frac{ft}{ft}}{0.02 \frac{ft}{ft}/\text{sec}}$$
$$= 3.5 \text{ sec}$$

The length of superelevation must be at least the minimum transition time multiplied by the design speed.

$$L_{min} = (\text{min transition time})(\text{speed})$$
$$= (3.5 \text{ sec})\left(102.7 \frac{ft}{sec}\right)$$
$$= 359.5 \text{ ft} \quad (360 \text{ ft})$$

The answer is (B).

Why Other Options Are Wrong

(A) This incorrect answer is the distance traveled in one second, which is an insufficient length to obtain the full superelevation.

(C) In this incorrect answer, transitioning the full 0.08 ft/ft and ignoring the pavement normal cross slope of 0.01 ft/ft results in too long of a transition.

(D) In this incorrect answer, including the reverse-slope runout in the transition length results in too long of a distance.

SOLUTION 69

The roadway is already sloped down to the right at 0.01 ft/ft, and the superelevation will increase the slope of the roadway down to the right, since this is a right-hand curve. The difference between normal slope and full superelevation is found by subtracting the normal slope from the full slope.

$$e_{\text{full}} - e_{\text{normal}} = 0.08 \, \frac{\text{ft}}{\text{ft}} - 0.01 \, \frac{\text{ft}}{\text{ft}}$$
$$= 0.07 \text{ ft/ft}$$

Determine the right-edge profile elevation change.

$$\Delta e = (\text{pavement width})(\text{slope change})$$
$$= (24 \text{ ft})\left(0.07 \, \frac{\text{ft}}{\text{ft}}\right)$$
$$= 1.68 \text{ ft}$$

Using the edge transition rate of 1:200, determine the transition length.

$$L = \Delta e(\text{change rate})$$
$$= (1.68 \text{ ft})\left(200 \, \frac{\text{ft}}{\text{ft}}\right)$$
$$= 336 \text{ ft} \quad (340 \text{ ft})$$

The answer is (B).

Why Other Options Are Wrong

(A) This incorrect answer is the transition rate, not the transition distance.

(C) In this incorrect answer, transitioning the full 0.08 ft/ft would result in too long of a transition.

(D) This incorrect answer results from improper interpretation of the transition-rate definition.

$$\frac{200 \text{ ft}}{0.07 \, \frac{\text{ft}}{\text{ft}}} = 2857 \text{ ft}$$

SOLUTION 70

Option I, curve radius, along with speed, determines the radial force needed to hold a vehicle on the curve.

Option II, passenger comfort, is important so that the driver can maintain control of the vehicle and so that passengers are not subject to unnecessary disorientation or the feeling of danger.

Option III, sight distance, is necessary so that the driver can avoid hazards.

Option VI, superelevation, allows the force of gravity to help keep the vehicle from sliding off the outside of the curve.

Weather conditions (option VIII), such as rain or snow, reduce the available tire friction needed to hold the vehicle on the curve.

The answer is (C).

Why Other Options Are Wrong

(A) This answer is incorrect. It omits weather conditions, such as rain or snow, which reduce the available tire friction needed to hold the vehicle on a curve. In regions that are often wet or icy, the design speed is set lower than in regions that have normally dry conditions.

(B) This answer is incorrect. It omits option II, passenger comfort, which is necessary so that the driver can maintain control of the vehicle and so that passengers are not subject to disorientation or a feeling of danger.

(D) This answer is incorrect. The posted speed limit, option V, is subject to local regulatory conditions and does not determine design speed of an existing highway. Posted speed can, however, be used to set the minimum design speed of a new or reconstructed highway.

SOLUTION 71

Find the curve tangent.

$$T = R \tan \frac{I}{2} = (1400 \text{ ft}) \tan \frac{32°20'}{2}$$
$$= 405.85 \text{ ft}$$

Determine the point of tangent (PT) coordinates using the tangent bearing and distance from the point of intersection (PI) coordinates.

$$N_{\text{PT}} = N_{\text{PI}} + T \sin(90° - \text{bearing angle})$$
$$= 5280 \text{ ft} + (405.85 \text{ ft}) \sin(90° - 55°40')$$
$$= 5508.90 \text{ ft} \quad (5510 \text{ ft})$$

The answer is (C).

Why Other Options Are Wrong

(A) This incorrect answer is the result of calculating the east coordinate instead of the north coordinate.

(B) This incorrect answer results from using the curve deflection angle to locate the point of tangent (PT) coordinates from the point of intersection (PI) instead of using the bearing of the ahead tangent.

(D) This incorrect answer results from neglecting to subtract the tangent bearing from 90° when determining the point of tangent (PT) coordinate.

SOLUTION 72

Determine the length of the back tangent (from the point of curvature (PC) to the point of intersection (PI)) by the difference of coordinates.

$$T = \sqrt{(N_{PI} - N_{PC})^2 + (E_{PI} - E_{PC})^2}$$
$$= \sqrt{(5280 \text{ ft} - 5280 \text{ ft})^2 (3546 \text{ ft} - 2640 \text{ ft})^2}$$
$$= 906 \text{ ft}$$

Find the curve deflection.

$$T = R \tan \frac{I}{2}$$

Rearrange to find I.

$$I = 2 \arctan \frac{T}{R}$$
$$= 2 \arctan \frac{906 \text{ ft}}{2300 \text{ ft}}$$
$$= 43°$$

Find the bearing of the ahead tangent (PI to point of tangent (PT)). The back tangent (line PC to PI) lies in a due-east position. Bearings are taken from due south or due north.

$$90° - 43° = 47°$$

The ahead tangent will be in the southeasterly quadrant. The bearing is S 47° E.

The answer is (C).

Why Other Options Are Wrong

(A) This incorrect answer is the azimuth of the ahead tangent.

(B) This incorrect answer is the result of using the curve deflection angle as the bearing angle.

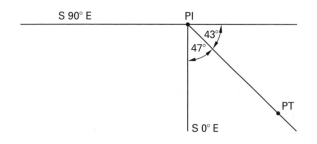

(D) This incorrect answer results from neglecting to double the angle obtained from the arctangent T/R.

SOLUTION 73

The basic relationship is

$$R_h = R_{rr} - 125 \text{ ft}$$

Convert the railroad curve to the radius definition using the chord definition formula.

$$R_{rr} = \frac{50}{\sin \dfrac{D}{2}} = \frac{50°\text{-ft}}{\sin \dfrac{6°}{2}}$$
$$= 955.37 \text{ ft}$$

Determine the highway centerline radius.

$$R_h = R_{rr} - 125.00 \text{ ft}$$
$$= 955.37 \text{ ft} - 125 \text{ ft}$$
$$= 830.37 \text{ ft}$$

Determine the highway degree of curve using the arc definition formula.

$$D = \frac{(180°)(100)}{\pi R} = \frac{(180°)(100 \text{ ft})}{\pi (830.37 \text{ ft})}$$
$$= 6.9000° \quad (6°54'00'')$$

The answer is (B).

Why Other Options Are Wrong

(A) This incorrect answer results from placing the railroad curve to the inside of the highway (i.e., adding 125 ft to the railroad radius).

(C) This incorrect answer results from using the arc definition for the railroad curve.

(D) This incorrect answer results from not dividing D by 2 in the chord definition equation.

SOLUTION 74

Superelevation, e, of a railroad track is the amount of rise (elevation) given to the outer rail of a curve at the gauge line over the inner rail. The transition length, s, is determined by dividing the superelevation of the curve, e, by the allowable rate of change in inches per second, Δe, times the velocity.

$$s = \left(\frac{e}{\Delta e}\right) v$$

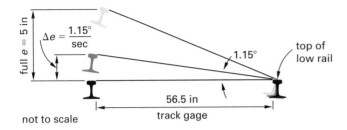

The rate-of-change angle can be converted to the rate of change in elevation using the tangent of the rate-of-change angle. The tangent of the rate-of-change cross-slope angle is the elevation change of the high rail, Δe, divided by the track gage, 56.5 in.

$$\frac{\tan 1.15°}{\sec} = 0.02$$

Determine the rate of change, Δe, in inches per second.

$$\Delta e = (0.02)(56.5 \text{ in})$$
$$= 1.13 \text{ in/sec}$$

Determine the number of seconds, t, to attain full superelevation.

$$t = \frac{e}{\Delta e} = \frac{5 \text{ in}}{1.13 \frac{\text{in}}{\text{sec}}}$$
$$= 4.42 \text{ sec}$$

Determine the distance traveled in 4.42 sec.

$$s = vt$$
$$= \left(150 \frac{\text{mi}}{\text{hr}}\right)\left(5280 \frac{\text{ft}}{\text{mi}}\right)\left(\frac{1 \text{ hr}}{3600 \text{ sec}}\right)(4.42 \text{ sec})$$
$$= 972 \text{ ft} \quad (970 \text{ ft})$$

The answer is (C).

Why Other Options Are Wrong

(A) This incorrect answer results from dividing the full superelevation by the tangent of the rate-of-change angle.

(B) This incorrect answer results from a faulty speed conversion.

(D) This incorrect answer results from dividing the speed by the tangent of the rate-of-change angle.

SOLUTION 75

The layout shows all mainline exits and entrances on the right side of the roadway.

The closeness of the inner entrance and exit ramps can make it more difficult to navigate through the weaving zone in heavy traffic. Accelerating entrance traffic will conflict with decelerating exit traffic.

The answer is (D).

Why Other Options Are Wrong

(A) This answer is incorrect. The one bridge needed is very wide and very long but not necessarily less costly than several smaller bridges.

(B) This answer is incorrect. Drivers may well understand the layout, but this interchange is more difficult to sign than are direct connection ramps. The large inner ramp loops consume much real estate that is land locked by the ramps; therefore, this interchange does not make efficient use of the land.

(C) This answer is incorrect. The only movements that are direct connections are the outer movements. Loop ramps are not considered direct connections. The smaller radius of the inner loop ramps requires considerable reduction in speed from the mainline.

SOLUTION 76

Referring to Hickerson's *Route Location and Design*, a formula for obtaining the external distance of the curve is

$$E = (R + p)\left(\sec \frac{I}{2}\right) - R$$

To find the secant,

$$\sec \frac{I}{2} = \frac{1}{\cos \frac{I}{2}} = \frac{1}{\cos \frac{24°}{2}}$$
$$= \frac{1}{0.978148}$$

To find p,

$$p = y - R(1 - \cos I_S)$$

To find y,

$$y = \frac{L_s^2}{6R} = \frac{(250 \text{ ft})^2}{(6)(850 \text{ ft})}$$
$$= 12.25 \text{ ft}$$

To find the spiral angle,

$$I_s = \left(\frac{L_s}{200}\right)\left(\frac{(180°)(100)}{\pi R}\right)$$
$$= \left(\frac{250 \text{ ft}}{200 \text{ ft}}\right)\left(\frac{(180°)(100 \text{ ft})}{\pi(850 \text{ ft})}\right)$$
$$= 8.42585°$$

Then,

$$p = 12.25 \text{ ft} - (850 \text{ ft})(1 - 0.989206)$$
$$= 3.08 \text{ ft}$$

$$E = (850 \text{ ft} + 3.08 \text{ ft})\left(\frac{1}{0.978148}\right) - 850 \text{ ft}$$
$$= 22.14 \text{ ft} \quad (22 \text{ ft})$$

The answer is (B).

Why Other Options Are Wrong

(A) In this incorrect answer, not including the p distance in the external distance equation yields an answer that is too small.

(B) In this incorrect answer, using the value for y instead of the value for p in the external distance equation yields an answer that is too large.

(D) In this incorrect answer, not dividing the deflection by two in the external distance equation yields an answer that is too large.

SOLUTION 77

Determine the curve radius.

$$R = \frac{(180°)(100)}{\pi D} = \frac{(180°)(100 \text{ ft})}{\pi 5.73°}$$
$$= 1000 \text{ ft}$$

Determine the spiral length using the given formula.

$$L_s = 1.6\left(\frac{v^3}{R}\right) = \left(1.6 \frac{\text{ft}^2\text{-hr}^3}{\text{mi}^3}\right)\left(\frac{\left(60 \frac{\text{mi}}{\text{hr}}\right)^3}{1000 \text{ ft}}\right)$$
$$= 346 \text{ ft} \quad (350 \text{ ft})$$

Determine the change in cross slope.

$$\text{change in cross slope} = \text{full slope} - \text{normal slope}$$
$$= 0.08 \frac{\text{ft}}{\text{ft}} - 0.015 \frac{\text{ft}}{\text{ft}}$$
$$= 0.065 \text{ ft/ft}$$

Determine the change in edge of pavement elevation for a three-lane half width of roadway.

$$\Delta e = ew$$
$$= \left(0.065 \frac{\text{ft}}{\text{ft}}\right)\left(12 \frac{\text{ft}}{\text{lane}}\right)(3 \text{ lanes})$$
$$= 2.34 \text{ ft}$$

Using the edge transition rate of 1:200, determine the transition length.

$$L_{\text{tr}} = (\Delta e) G_r$$
$$= (2.34 \text{ ft})\left(200 \frac{\text{ft}}{\text{ft}}\right)$$
$$= 468 \text{ ft}$$

The slope ratio transition length, 468 ft, is greater than the spiral formula length, 346 ft. Using the longer of the two criteria, set the spiral length to 470 ft.

The answer is (B).

Why Other Options Are Wrong

(A) This incorrect answer is the spiral length using the spiral formula, which is shorter than the edge transition criterion.

(C) This incorrect answer is the result of using the full slope transition without considering that the pavement is already sloped 0.015 ft/ft at the beginning of the transition.

(D) This incorrect answer results from using the entire six-lane width to determine the edge transition elevation.

SOLUTION 78

The design speed determines the distance necessary to change from one curve radius to the next, which gives the driver time to adjust to the new radius. The formula given in the problem for spiral length is the GDHS formula, which is based on speed in miles per hour and radius in feet when the change in lateral acceleration is 2.0 ft/sec². The formula is for the length of a spiral necessary to transition from the tangent to the radius of a central curve in which the radius changes in proportion to the distance along the spiral. The length of spiral required is the length necessary to transition from the equivalent radius of a 3° curve to a 2° curve. The radius of a curve is determined by

$$R = \frac{(180°)(100)}{\pi D}$$
$$= \frac{(180°)(100 \text{ ft})}{\pi D}$$

For each curve,

$$R_1 = \frac{(180°)(100 \text{ ft})}{\pi 3°}$$
$$= 1909.86 \text{ ft}$$

$$R_2 = \frac{(180°)(100 \text{ ft})}{\pi 2°}$$
$$= 2864.79 \text{ ft}$$

The length of the transition is determined by finding the distance along the spiral in which the radius changes from R_1 to R_2.

$$L_s = L_1 - L_2 = \frac{1.6v^3}{R_1} - \frac{1.6v^3}{R_2}$$

$$= \left(\frac{\left(1.6 \frac{\text{ft}^2\text{-hr}^3}{\text{mi}^3}\right)\left(70 \frac{\text{mi}}{\text{hr}}\right)^3}{1909.53 \text{ ft}} \right)$$

$$- \left(\frac{\left(1.6 \frac{\text{ft}^2\text{-hr}^3}{\text{mi}^3}\right)\left(70 \frac{\text{mi}}{\text{hr}}\right)^3}{2864.79 \text{ ft}} \right)$$

$$= 95.78 \quad (100 \text{ ft})$$

The answer is (A).

Why Other Options Are Wrong

(B) This incorrect answer is the length of spiral needed at the tangent end of a 2° curve.

(C) This incorrect answer is the length of spiral needed at the tangent end of a 3° curve.

(D) This incorrect answer is the length of spiral needed if a full spiral were introduced between the curves that transitioned to tangent and then transitioned into the second curve. This method is not a true compound spiral curve but rather two fully spiraled curves back-to-back.

SOLUTION 79

The formula given for the spiral length in the problem is the GDHS formula based on speed in miles per hour and radius in feet, when the change in lateral acceleration is 2.0 ft/sec². The curve radius calculation using degree of curve can be incorporated into the base equation by substituting for R.

$$R = \frac{(360°)(100)}{2\pi D}$$

$$L_s = 1.6v^3 \left(\frac{2\pi D}{(360°)(100)} \right)$$

The total length of transition is determined by the total change in degree of curve. Determine the change in degree of curve.

$$D_1 - D_2 = 2° - (-3)$$
$$= 5°$$

Determine the length of the transition spiral required.

$$L_s = (1.6)\left(60 \frac{\text{mi}}{\text{hr}}\right)^3 \left(\frac{2\pi(5°)}{(360°)(100 \text{ ft})} \right)$$

$$= 301.6 \text{ ft} \quad (300 \text{ ft})$$

The answer is (D).

Why Other Options Are Wrong

(A) This incorrect answer is the required length of spiral for a 1° change in degree of curve. This solution would be true for a compound curve but is incorrect for a reverse curve.

(B) This incorrect answer is the required length of spiral to bring a 2° curve back to zero deflection. More spiral length is required to develop the 3° curve in the opposite direction.

(C) This incorrect answer is the required length of spiral to develop a 3° curve from zero deflection (tangent). More spiral length is required to bring the 2° curve deflection to zero.

SOLUTION 80

The spiral offset is defined by

$$p = y_s - R \text{ vers } I_s$$

Determine the curve radius.

$$R = \frac{(180°)(100)}{\pi D} = \frac{(180°)(100 \text{ ft})}{\pi 5°}$$

$$= 1145.92 \text{ ft}$$

Determine the spiral angle.

$$I_s = \left(\frac{L_s}{200} \right) D = \left(\frac{200 \text{ ft}}{200 \text{ ft}} \right)(5°)$$

$$= 5°$$

Determine the tangent offset.

$$y_s = \frac{L_s^2}{6R} = \frac{(200 \text{ ft})^2}{(6)(1145.92 \text{ ft})}$$

$$= 5.82 \text{ ft}$$

Determine the spiral offset.

$$p = 5.82 \text{ ft} - (1145.92 \text{ ft}) \text{vers } 5°$$

$$= 1.46 \text{ ft}$$

The answer is (B).

Why Other Options Are Wrong

(A) This incorrect value results from using the sine of the spiral deflection instead of the versine in the spiral offset equation.

(C) This incorrect answer is the external distance of the full curve, which indicates how much the point of intersection (PI) is shifted radially from the unspiraled curve.

(D) This incorrect answer is the tangent offset, y_s, of the full spiral at the spiral-to-curve point (SC).

SOLUTION 81

Assume $S < L$. From GDHS Exh. 3-76, the stopping sight distance for a 50 mph design speed is 425 ft. Formulas 3-45 and 3-47 are used when the height of eye is 3.5 ft and the height of object is 2.0 ft.

$$L = \frac{AS^2}{2158} = \frac{(3\% - (-4\%))(425 \text{ ft})^2}{2158 \text{ ft}}$$
$$= 588 \text{ ft} \quad (590 \text{ ft})$$

Assume $S > L$.

$$L = 2S - \frac{2158}{A} = (2)(425 \text{ ft}) - \frac{2158 \text{ ft}}{(3\% - (-4\%))}$$
$$= 542 \text{ ft} \quad (540 \text{ ft})$$

Since 542 ft is greater than the stopping sight distance required, the first assumption—that the stopping sight distance is less than the curve length—is valid. The length can be checked using the minimum design K value from GDHS Exh. 3-76.

$$L = KA = \left(84 \frac{\text{ft}}{\%}\right)(7\%)$$
$$= 588 \text{ ft} \quad (590 \text{ ft})$$

The minimum required vertical curve length is 588 ft (590 ft).

The answer is (C).

Why Other Options Are Wrong

(A) This incorrect answer is the sight distance. It is not correct to assume that the curve length can be the same as the sight distance. A curve of this length would have a stopping sight distance of 360 ft, which is for a speed of 45 mph.

(B) Even though 540 ft would allow a shorter curve than the correct answer, the assumption that the sight distance is longer than the curve length is incorrect, making this curve length inadequate.

(D) Using an object height of 0.5 ft (from previous GDHS criteria) yields a curve length that is greater than the minimum by GDHS 2000 criteria, as in this incorrect answer.

SOLUTION 82

Check for the approximate required sight distance from GDHS Exh. 3-1. The recommended stopping distance for 70 mph is 730 ft. Therefore, the curve length of 1500 ft is probably greater than the required stopping sight distance.

Check the available sight distance. Use the GDHS formula for crest vertical curves for stopping sight distances less than the curve length.

$$L = \frac{AS^2}{2158} = \frac{(G_2 - G_1)S^2}{2158 \text{ ft}}$$
$$= 1500 \text{ ft}$$

Solve for the sight distance available.

$$S = \sqrt{\frac{2158L}{A}} = \sqrt{\frac{(2158 \text{ ft})(1500 \text{ ft})}{3.0\% - (-2.5\%)}}$$
$$= 767 \text{ ft}$$

Examining GDHS Exh. 3-3, 767 ft appears in the lower range of the stopping sight distance for 50 mph conditions. However, designing for minimum stopping sight distance under wet conditions, which is the normal GDHS design criteria, is not entirely safe. In fog conditions, a car braking rapidly or coming to a full stop can cause multiple rear-end collisions. Therefore, avoidance maneuvers A and B do not apply, and one of the remaining conditions, C, D, or E, applies. Avoidance maneuvers C, D, and E allow for a greater perception-reaction time and allow time for the driver to change vehicle path or speed when a full stop situation is undesirable. Avoidance maneuver C at 50 mph requires 750 ft, which is less than the 767 ft available. Avoidance maneuver D allows 40 mph maximum speed and avoidance maneuver E allows 35 mph maximum for the available 767 ft sight distance. Therefore, for any avoidance maneuver other than stopping conditions, the speed limit should be posted at 50 mph maximum.

The answer is (A).

Why Other Options Are Wrong

(B) In this incorrect answer, posting for 60 mph places the 767 ft stopping sight distance within the range of 610 ft to 1150 ft for avoidance maneuvers A and B, which does not leave the additional distance required for lane change decision and adverse weather conditions.

(C) In this incorrect answer, misinterpreting GDHS Exh. 3-2 by using the former criteria of a 0.6 ft object height yields a longer stopping sight distance required than does using the newer criteria. Applying the value to GDHS Exh. 3-2 yields a speed limit of 60 mph.

(D) In this incorrect answer, using the passing sight distance from GDHS Exh. 3-7 yields a condition that will be in the mid to upper range of 70 mph values shown in GDHS Exhs. 3-1 and 3-2. This leads to selection of 70 mph for the answer.

SOLUTION 83

Determine the stopping sight distance required for a design speed of 35 mph using GDHS Exh. 3-1. The recommended design distance is 250 ft.

Assume the curve length will be longer than the required sight distance.

$$L = \frac{AS^2}{400 + 3.5S} = \frac{(G_2 - G_1)(250 \text{ ft})^2}{400 \text{ ft} + (3.5)(250 \text{ ft})}$$

$$= \frac{\left(3.5 \frac{\text{ft}}{\text{ft}} - \left(-6 \frac{\text{ft}}{\text{ft}}\right)\right)(250 \text{ ft})^2}{400 \text{ ft} + (3.5)(250 \text{ ft})}$$

$$= 466 \text{ ft}$$

Verify by using the formula for a curve length that is shorter than the required sight distance.

$$L = 2S - \frac{400 + 3.5S}{A}$$

$$= 2S - \frac{400 \text{ ft} + (3.5)(250 \text{ ft})}{G_2 - G_1}$$

$$= (2)(250 \text{ ft}) - \frac{400 \text{ ft} + (3.5)(250 \text{ ft})}{3.5 \frac{\text{ft}}{\text{ft}} - \left(-6.0 \frac{\text{ft}}{\text{ft}}\right)}$$

$$= 366 \text{ ft} \quad (370 \text{ ft})$$

Both answers are greater than the required 250 ft. Therefore, the required length is 466 ft (500 ft).

The answer is (C).

Why Other Options Are Wrong

(A) In this incorrect answer, determining A to be the arithmetic instead of algebraic difference in grades results in too short of a curve length using the $S < L$ formula.

(B) In this incorrect answer, using the stopping sight distance of 250 ft for the length of vertical curve results in too short of a vertical curve.

(D) In this incorrect answer, determining A to be the arithmetic instead of algebraic difference in grades results in much too long of a curve length using the $S > L$ formula.

SOLUTION 84

Sight distance around an obstruction is the distance along the centerline of the traveled lane, which is on a 250 ft radius. The clearance distance is the distance from the lane edge plus one-half of the lane width.

$$Z_{\text{clearance}} = \text{edge clearance} + \frac{\text{lane width}}{2}$$

$$= 14 \text{ ft} + \frac{12 \text{ ft}}{2}$$

$$= 20 \text{ ft}$$

Use the clearance distance, $Z_{\text{clearance}}$, as the mid-ordinate, M, of a circular curve arc centered at the point of clearance. Determine the curve length using the mid-ordinate and the radius.

$$M = R\left(1 - \cos\frac{I}{2}\right)$$

The deflection, I, is the angle subtended by the ends of the sight-line chord. Rearranging and solving for I,

$$\frac{M}{R} = \frac{20 \text{ ft}}{250 \text{ ft}}$$

$$= 1 - \cos\frac{I}{2}$$

$$I = 46.148°$$

The length of curve, L, can be determined using the radius and the degree of curve, D.

$$L = \frac{100I}{D} = \frac{\pi I R}{180°}$$

$$= \frac{\pi(46.148°)(250 \text{ ft})}{180°}$$

$$= 201.4 \text{ ft}$$

A sight distance of 201.4 ft is good for a 30 mph design speed.

The answer is (C).

Alternate Solution:

Using GDHS Exh. 3-57, from a point at 20 ft mid-ordinate on the bottom horizontal scale, intersect a line from a point at a 250 ft radius on the left vertical scale. The intersection lies on the v = 30 mph curve.

Why Other Options Are Wrong

(A) After determining $I/2$ from the mid-ordinate equation, this incorrect answer results from not doubling the answer for the length-of-curve equation.

(B) This answer results from incorrectly using the radius of the inside edge of the lane as the sight line.

(D) This answer results from incorrectly using the radius of the outside edge of the traveled lane as the sight line.

TRAFFIC SAFETY

SOLUTION 85

The deceleration equation has two unknowns, the initial velocity, v_0, and the friction factor, f, which can be solved by trial and error.

$$s = \frac{v_0^2}{2g(f+G)} = \frac{v_0^2}{(2)\left(32.2 \frac{\text{ft}}{\text{sec}^2}\right)(f+G)}$$

Assume the initial speed is 40 mph. Therefore, the friction factor is 0.51 for the first trial.

$$v_0 = \sqrt{(\text{skid distance})(2g)(f+G)}$$
$$= \sqrt{(185 \text{ ft})\left((2)\left(32.2 \frac{\text{ft}}{\text{sec}^2}\right)\right)\left(0.51 - 0.03 \frac{\text{ft}}{\text{ft}}\right)}$$
$$= 75.6 \text{ ft/sec}$$

Convert to miles per hour.

$$v_0 = \left(75.6 \frac{\text{ft}}{\text{sec}}\right)\left(\frac{1 \text{ mi}}{5280 \text{ ft}}\right)\left(3600 \frac{\text{sec}}{\text{hr}}\right)$$
$$= 51.6 \text{ mph}$$

This result is faster than 40 mph.

Try an initial speed of 50 mph with a friction factor of 0.45.

$$v_0 = \sqrt{(185 \text{ ft})\left((2)\left(32.2 \frac{\text{ft}}{\text{sec}^2}\right)\right)\left(0.45 - 0.03 \frac{\text{ft}}{\text{ft}}\right)}$$
$$= 70.7 \text{ ft/sec}$$

Convert to miles per hour.

$$v_0 = \left(70.7 \frac{\text{ft}}{\text{sec}}\right)\left(\frac{1 \text{ mi}}{5280 \text{ ft}}\right)\left(3600 \frac{\text{sec}}{\text{hr}}\right)$$
$$= 48.2 \text{ mph}$$

This result is less than 50 mph. Therefore, the initial speed must be between 40 mph and 50 mph. Try an initial speed of 49 mph.

Determine the friction factor by linear interpolation.

$$f = \frac{0.51 - 0.45}{10} + 0.45$$
$$= 0.456$$

Use this value to determine the initial speed.

$$v_0 = \sqrt{(185 \text{ ft})\left((2)\left(32.2 \frac{\text{ft}}{\text{sec}^2}\right)\right)\left(0.456 - 0.03 \frac{\text{ft}}{\text{ft}}\right)}$$
$$= 71.24 \text{ ft/sec}$$

Convert to miles per hour.

$$v_0 = \left(71.24 \frac{\text{ft}}{\text{sec}}\right)\left(\frac{1 \text{ mi}}{5280 \text{ ft}}\right)\left(3600 \frac{\text{sec}}{\text{hr}}\right)$$
$$= 48.6 \text{ mph} \quad (49 \text{ mph})$$

The answer is (B).

Why Other Options Are Wrong

(A) This incorrect answer results from improper rationalizing of the equation throughout the trial-and-error process.

(C) This incorrect answer results from ignoring the downgrade of the skid.

(D) This incorrect answer is the initial speed in feet per second, not miles per hour.

SOLUTION 86

Use the formula that relates braking distance to speed and deceleration, considering initial and final velocities.

$$s = \frac{v_1^2 - v_2^2}{2d}$$

The deceleraton rate is determined by the braking force, which is limited by the friction factor, f. Deceleration is increased or decreased by the effect of gravity on the grade. Rewrite the formula, replacing deceleration with the relationship for friction, grade, and gravity.

$$s_b = \frac{v_1^2 - v_2^2}{2g(f+G)}$$

The formula can be written in the simplified form shown in GDHS to include conversions for velocity in miles per hour and distance in feet.

$$s_b = \frac{(v_1^2 - v_2^2)\left(\left(5280 \frac{\text{ft}}{\text{mi}}\right)\left(\frac{1 \text{ hr}}{3600 \text{ sec}}\right)\right)^2}{2\left(32.2 \frac{\text{ft}}{\text{sec}^2}\right)(f+G)}$$

$$= \frac{v_1^2 - v_2^2}{\left(30 \frac{\text{mi}^2}{\text{hr}^2\text{-ft}}\right)(f+G)}$$

The grade is negative since the car skids downhill. This has the effect of reducing the speed for a given length of the skid.

Rearrange the equation and solve for v_2.

$$v_2 = \sqrt{v_1^2 - s_b 30(f+G)}$$

$$= \sqrt{\left(70 \, \frac{\text{mi}}{\text{hr}}\right)^2 - (350 \text{ ft})\left(30 \, \frac{\text{mi}^2}{\text{hr}^2\text{-ft}}\right)\left(0.30 - 0.05 \, \frac{\text{ft}}{\text{ft}}\right)}$$

$$= 47.7 \text{ mph} \quad (48 \text{ mph})$$

The answer is (C).

Why Other Options Are Wrong

(A) This incorrect answer results from adding the grade to the friction factor instead of subtracting.

(B) This incorrect answer results from not including the grade but considering tire friction only.

(D) This incorrect answer results from assuming v_2 is 0 mph and reducing the form to the equation to solve for v.

SOLUTION 87

The equation for stopping distance, relating speed and deceleration rate, can be written in the simplified form shown in the GDHS to include conversions for velocity in miles per hour and distance in feet.

$$s_b = \frac{v_0^2}{2g(f+g)}$$

$$= \frac{v_0^2 \left(\left(5280 \, \frac{\text{ft}}{\text{mi}}\right)\left(\frac{1 \text{ hr}}{3600 \text{ sec}}\right)\right)^2}{2\left(32.2 \, \frac{\text{ft}}{\text{sec}^2}\right)(f+G)}$$

$$= \frac{v_0^2}{\left(30 \, \frac{\text{mi}^2}{\text{hr}^2\text{-ft}}\right)(f+G)}$$

There is only one term for velocity because the final speed is 0 mph. The term f is the friction factor and G is the grade in decimal.

Solve for the total braking distance from 60 mph if the car did not stike the concrete barriers.

$$s_b = \frac{\left(60 \, \frac{\text{mi}}{\text{hr}}\right)^2}{\left(30 \, \frac{\text{mi}^2}{\text{hr}^2\text{-ft}}\right)\left(0.30 + 0.03 \, \frac{\text{ft}}{\text{ft}}\right)}$$

$$= 364 \text{ ft}$$

The car already skidded 150 ft before the impact. If the concrete barriers were not encountered, the additional distance to stop would be

$$D_b = s_b - \text{skid distance before impact}$$
$$= 364 \text{ ft} - 150 \text{ ft}$$
$$= 214 \text{ ft} \quad (210 \text{ ft})$$

The answer is (B).

Why Other Options Are Wrong

(A) This incorrect answer results from improper conversion between miles per hour and feet per second.

(C) This incorrect answer results from ignoring the grade.

(D) This incorrect answer results from subtracting the grade from the friction factor instead of adding it to the friction factor.

SOLUTION 88

The initial speed is determined by the deceleration formula, relating change in speed to the braking distance and stopping rate.

$$s_b = \frac{v_2^2 - v_1^2}{2a}$$

Since the initial speed at the start of the skid is unknown, rearrange the equation to place the initial speed on the left.

$$v_1 = \sqrt{v_2^2 - 2as_b}$$

The acceleration (or deceleration) rate is determined by the friction factor and the grade.

$$a = g(f+G)$$
$$= \left(9.81 \, \frac{\text{m}}{\text{s}^2}\right)\left(0.48 + (-0.06 \, \frac{\text{m}}{\text{m}})\right)$$
$$= 4.12 \text{ m/s}^2$$

Since the situation described is a deceleration situation, the acceleration will be negative when solving the initial equation.

$$v_1 = \sqrt{\left(\left(40 \, \frac{\text{km}}{\text{h}}\right)\left(1000 \, \frac{\text{m}}{\text{km}}\right)\left(\frac{1 \text{ h}}{3600 \text{ s}}\right)\right)^2 - (2)\left(-4.12 \, \frac{\text{m}}{\text{s}^2}\right)(60 \text{ m})}$$

$$= 24.9 \text{ m/s}$$

Convert to kilometers per hour.

$$v_1 = \left(24.9 \, \frac{\text{m}}{\text{s}}\right)\left(\frac{1 \text{ km}}{1000 \text{ m}}\right)\left(3600 \, \frac{\text{s}}{\text{h}}\right)$$
$$= 89.6 \text{ kph} \quad (90 \text{ kph})$$

The answer is (C).

Why Other Options Are Wrong

(A) This incorrect answer is the initial speed in miles per second, not kilometers per hour.

(B) This incorrect answer results from not considering acceleration as negative and then ignoring the negative sign under the square root.

(D) This incorrect answer results from not including the effect of grade on deceleration.

SOLUTION 89

For the first part of the skid, use the formula that relates the change in distance to the difference in the square of the speeds divided by the deceleration rate. Use the form of the equation where speed is in miles per hour and skid distance is in feet.

$$s_b = \frac{v_1^2 - v_0^2}{30(f+G)}$$

For this formula, speed is in miles per hour, skid distance is in feet, and the conversion constant 30 has units of mi^2/hr^2-ft and is negative for deceleration. The friction factor, f, and the grade, G, are in decimal form.

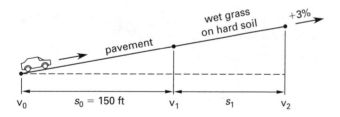

Rearrange the formula to isolate initial speed on the left side.

$$v_1 = \sqrt{v_0^2 + (-30(f+G))\,s_0}$$

$$= \sqrt{\left(60\,\frac{mi}{hr}\right)^2 + \left(\left(-30\,\frac{mi^2}{hr^2\text{-ft}}\right)\left(0.30 + 0.03\,\frac{ft}{ft}\right)\right) \times (150\,\text{ft})}$$

$$= 46\,\text{mph}$$

Reapply the initial formula to determine the skid distance required to stop from 46 mph on grass.

$$s_1 = \frac{v_2^2 - v_1^2}{-30(f+G)}$$

$$= \frac{(0)^2 - \left(46\,\frac{mi}{hr}\right)^2}{\left(-30\,\frac{mi^2}{hr^2\text{-ft}}\right)\left(0.10 + 0.03\,\frac{ft}{ft}\right)}$$

$$= 543\,\text{ft} \quad (540\,\text{ft})$$

The answer is (A).

Why Other Options Are Wrong

(B) This incorrect answer results from not including the effect of the grade on the stopping distance.

(C) This incorrect answer results from not using a negative conversion constant, -30, to denote deceleration.

(D) This incorrect answer results from inserting the speed value in feet per second in place of the speed value in miles per hour.

SOLUTION 90

For a segment of highway, the accident rates are reported per HMVM traveled per year. The formula is

$$R = \frac{(\text{no. of injury accidents})(10^8)}{(\text{ADT})(\text{no. of sample years})\left(365\,\frac{\text{days}}{\text{yr}}\right)L}$$

$$\text{total fatal accidents} = 0 + 3 + 2 + 5 = 10 \text{ accidents}$$

$$\text{total personal injury accidents} = 35 + 35 + 50 + 50$$
$$= 170 \text{ accidents}$$

Therefore,

$$\text{total injury accidents} = 10\text{ fatal} + 170\text{ personal injury}$$
$$= 180 \text{ accidents}$$

The average daily traffic (ADT) is

$$\text{ADT}_{\text{ave}} = \frac{\sum \text{ADT}}{\text{no. of sample years}}$$

$$= \frac{15{,}500\,\frac{\text{veh}}{\text{day}} + 16{,}000\,\frac{\text{veh}}{\text{day}} + 16{,}500\,\frac{\text{veh}}{\text{day}} + 17{,}000\,\frac{\text{veh}}{\text{day}}}{4}$$

$$= 16{,}250 \text{ veh/day}$$

Determine the injury accident rate.

$$R = \frac{(180 \text{ accidents})\left(10^8 \frac{\text{veh-mi}}{\text{HMVM}}\right)}{\left(16{,}250 \frac{\text{veh}}{\text{day}}\right)(4 \text{ yr})\left(365 \frac{\text{days}}{\text{yr}}\right)(20 \text{ mi})}$$

$$= 37.93 \text{ accidents/HMVM} \quad (38 \text{ accidents/HMVM})$$

The answer is (C).

Why Other Options Are Wrong

(A) This incorrect answer is the injury rate per million annual vehicle miles instead of per hundred million annual vehicle miles.

(B) This incorrect answer results from erroneously counting only the personal injury accidents in the rate, neglecting to include the fatal accidents.

(D) This incorrect answer results from erroneously counting all accidents in the rate, including property damage only.

SOLUTION 91

Determine the traffic base.

$$\text{traffic base} = (\text{ADT})L$$
$$= \left(28{,}000 \frac{\text{veh}}{\text{day}}\right)\left(365 \frac{\text{days}}{\text{yr}}\right)(2.5 \text{ mi})$$
$$\times \left(10^8 \frac{\text{HMVM}}{\text{veh-mi}}\right)$$
$$= 0.256 \text{ HMVM/yr}$$

The critical rate for the statewide average accident rate is

$$R_{\text{crit}} = R_{\text{ave}} + K\sqrt{\frac{R_{\text{ave}}}{\text{traffic base}}}$$

$$= 150 \frac{\text{accidents}}{\text{HMVM}} + 1.645\sqrt{\frac{150 \frac{\text{accidents}}{\text{HMVM}}}{0.256 \text{ HMVM}}}$$

$$= 190 \text{ accidents/HMVM}$$

The accident rate for the particular highway segment being evaluated is

$$R_{\text{seg}} = \frac{\text{annual accident rate}}{\text{traffic base}}$$

$$= \frac{9 \frac{\text{accidents}}{\text{yr}}}{0.256 \frac{\text{HMVM}}{\text{yr}}}$$

$$= 35.2 \text{ accidents/HMVM}$$

Determine the ratio of the segment accident rate with respect to the statewide critical rate.

$$\frac{R_{\text{seg}}}{R_{\text{crit}}} = \frac{35.2 \frac{\text{accidents}}{\text{HMVM}}}{190 \frac{\text{accidents}}{\text{HMVM}}}$$

$$= 0.185 \quad (0.19)$$

The answer is (B).

Why Other Options Are Wrong

(A) This incorrect answer results from applying the test factor directly to the statewide accident rate.

(C) This answer results from incorrectly determining the critical rate.

(D) This incorrect answer results from comparing the statewide critical rate to the segment accident rate.

SOLUTION 92

Use the equation that finds deceleration distance by relating speed and deceleration rate.

$$s = \frac{v_{\text{final}}^2 - v_{\text{impact}}^2}{2a}$$

The final speed is 0 mph and the deceleration rate, or negative acceleration, is set by the number of force g's. The efficiency factor serves to add a margin of safety to the crash cushion by increasing its length over the minimum necessary. Insert these values in the equation and determine the final answer.

$$s_d = \frac{v_{\text{final}}^2 - v_{\text{impact}}^2}{2g\left(-\frac{a}{g}\right)\eta}$$

$$= \frac{\left(0 \frac{\text{mi}}{\text{hr}}\right)^2 - \left(\left(70 \frac{\text{mi}}{\text{hr}}\right)\left(5280 \frac{\text{ft}}{\text{mi}}\right)\left(\frac{1 \text{ hr}}{3600 \text{ sec}}\right)\right)^2}{(2)\left(32.2 \frac{\text{ft}}{\text{sec}^2}\right)(-7)(0.75)}$$

$$= 31 \text{ ft}$$

The answer is (D).

Why Other Options Are Wrong

(A) This incorrect answer results from an improper conversion from miles per hour to feet per second.

(B) This incorrect answer results from inversely applying the efficiency factor.

(C) This incorrect answer results from ignoring the efficiency factor in the equation.

SOLUTION 93

The deceleration is derived from Newton's equation, with the factor of safety, FS, applied.

$$P = M \left(\frac{a}{g_c}\right)(\text{FS})$$

Determine the acceleration (or deceleration) rate using the initial speed and the stopping distance. The formula is

$$s = \frac{v_2^2 - v_1^2}{2a}$$

Rearranging to find a,

$$a = \frac{v_1^2 - v_0^2}{2s}$$

$$= \frac{\left(0 \, \frac{\text{mi}}{\text{hr}}\right)^2 - \left(\left(70 \, \frac{\text{mi}}{\text{hr}}\right)\left(5280 \, \frac{\text{ft}}{\text{mi}}\right)\left(\frac{1 \, \text{hr}}{3600 \, \text{sec}}\right)\right)^2}{(2)(23.4 \, \text{ft})}$$

$$= -225.2 \, \text{ft/sec}^2$$

Normalize to the standard gravitational acceleration rate to check for the maximum.

$$\frac{-225.2 \, \frac{\text{ft}}{\text{sec}^2}}{\left(32.2 \, \frac{\text{ft}}{\text{sec}^2}\right)\left(\frac{1}{g}\right)} = -6.99g < 7.5g \quad [\text{OK}]$$

Determine the force on the backwall.

$$P = (4500 \, \text{lbm})\left(\frac{6.99g}{g_c}\right)(1.5)$$

$$= (4500 \, \text{lbm}) \left(\frac{(6.99)\left(32.2 \, \frac{\text{ft}}{\text{sec}^2}\right)}{32.2 \, \frac{\text{lbm-ft}}{\text{lbf-sec}^2}}\right)(1.5)$$

$$= 47{,}183 \, \text{lbf} \quad (47 \, \text{kips})$$

The answer is (B).

Why Other Options Are Wrong

(A) This incorrect answer does not apply the factor of safety to the deceleration force.

(C) This incorrect answer uses $7.5g$ as the deceleration, whereas the problem called for the deceleration rate resulting from a 23.4 ft stopping distance.

(D) This incorrect answer is the result of not using the gravitational constant, g_c, to achieve consistent units between mass and force.

SOLUTION 94

Left-turning vehicles require sharing of space with oncoming traffic flow. Large straight-through flow volumes combined with large left-turn flow volumes compete for time in the conflict zone shared by both flows.

Adequate visibility, aided by roadway lighting, is important, but is of little advantage if there are inadequate gaps in the flow to allow left-turning vehicles to cross the opposing flow.

If the intersection is signalized, a special left-turning phase can create needed gaps in traffic flow and would be more important than additional roadway lighting in most cases.

The answer is (B).

Why Other Options Are Wrong

(A) This answer is incorrect. A large volume of left turns is the most likely cause of head-on collisions.

(C) This answer is incorrect. The absence of a left-turning phase would generally be more important than the need for additional roadway lighting.

(D) This answer is incorrect. Inadequate gaps in the opposing flow of traffic encourage left-turn drivers to take more risks.

SOLUTION 95

Slippery road conditions require drivers to allow more spacing between vehicles than is normal for a given speed. Reducing the approach speed limit would encourage drivers to compensate for the poor surface, with little effect on traffic volume. Visibility improvements would naturally occur by having the traffic approach the intersection at a lower speed, which in turn would allow more time for reacting to unusual conditions at the intersection.

The answer is (A).

Why Other Options Are Wrong

(B) Overhead signals may or may not be warranted. Unless there is already a serious restriction on visibility of signals by approaching traffic, additional signals would probably have a negligible effect on conflicts caused by reduced traction.

(C) Improved roadway lighting would possibly improve visibility at night, but would have a negligible effect on conflicts caused by reduced traction.

(D) Prohibiting turns may eliminate some accidents, depending on traffic volumes and movement patterns. There would be a negligible effect on conflicts caused by reduced traction.

SOLUTION 96

While all of the countermeasures listed could lead to accident injury reduction, only one is not directly involved with highway design, construction, operation, or maintenance. That is the ergonomic design of vehicle interiors, which is covered by other agencies and industry standards.

The answer is (C).

Why Other Options Are Wrong

(A) This answer is incorrect. Signing, marking, and delineation are an integral part of on-highway safety improvements.

(B) This answer is incorrect. Enforcement of traffic laws, in conjunction with signing and marking programs, is a function of highway traffic management responsibility.

(D) This answer is incorrect. Pedestrians use roads, streets, and highways and must be accommodated in a safe fashion. Pedestrian management is a function of street and highway management responsibility.

SOLUTION 97

Determine the undepreciated amount at the end of the tenth year, starting with the total depreciation.

$$\frac{\text{total}}{\text{depreciation}} = \text{installation cost} - \text{salvage value}$$
$$= \$120{,}000 - \$15{,}000$$
$$= \$105{,}000$$

$$\frac{\text{depreciation}}{\text{per year}} = \frac{\text{total depreciation}}{\text{lifetime}}$$
$$= \frac{\$105{,}000}{15 \text{ yr}}$$
$$= \$7000/\text{yr}$$

$$\frac{\text{undepreciated}}{\text{value after 10 yr}} = \frac{\text{installation}}{\text{cost}} - (10 \text{ yr})\left(\frac{\text{depreciation}}{\text{per year}}\right)$$
$$= \$120{,}000 - (10 \text{ yr})\left(\frac{\$7000}{\text{yr}}\right)$$
$$= \$50{,}000$$

The tenth-year salvage value will be the present value of the future $15,000, at 6% per year.

$$P = F\left(\frac{P}{F}, i, n\right)$$

From factor tables for $i = 6\%$ and $n = 5$,

$$P = (\$15{,}000)(0.7473) = \$11{,}210$$

The remaining value is

$$\begin{aligned}\text{value after} \\ 10 \text{ yr}\end{aligned} = P + \begin{aligned}\text{undepreciated value} \\ \text{after 10 yr}\end{aligned}$$
$$= \$11{,}210 + \$50{,}000$$
$$= \$61{,}210 \quad (\$61{,}000)$$

The answer is (C).

Alternate Solution:

Using the mathematical formula,

$$P = F\left(\frac{1}{(1+i)^n}\right)$$
$$= (\$15{,}000)\left(\frac{1}{(1.06)^5}\right)$$
$$= \$11{,}210$$

Why Other Options Are Wrong

(A) This incorrect answer results from using the capital recovery method of determining the annual capital amount.

(B) This answer is incorrect because depreciation was taken to zero at the end of 15 yr, instead of to the salvage value.

(D) This answer is incorrect because the salvage value was taken as an increased value at year 10, assuming that the salvage value followed the interest rate back from the end-of-life amount to become a reversed increase in value.

$$S_{10} = S_{15}\left(\frac{F}{P}, i, n\right)$$
$$= (\$15{,}000)(1.3382)$$
$$= \$20{,}073$$

The remaining value is

$$\text{value after 10 yr} = P + \text{undepreciated value after 10 yr}$$
$$= \$20{,}073 + \$50{,}000$$
$$= \$70{,}073 \quad (\$70{,}000)$$

SOLUTION 98

The problem is looking for a future value of a present amount, compounded over 15 annual periods. The formula is

$$P(1+i)^n = F$$

Determine the future amount.

$$F = P(1+i)^n$$
$$= (\$2,500,000)(1+0.045)^{15}$$
$$= \$4,838,200 \quad (\$4,840,000)$$

The answer is (D).

Why Other Options Are Wrong

(A) This incorrect answer is the amount of interest gained, not the total value of the fund.

(B) This incorrect answer results from adding simple interest to the initial amount instead of compounding interest.

(C) This incorrect answer results from reversing the interest and term values.

SOLUTION 99

The annual cost is equal to the sum of annual bond interest plus the annual sinking fund amount. The formula for a sinking fund is

$$A = F\left(\frac{i}{(1+i)^n - 1}\right)$$

Determine the total amount required for the bond issue.

$$\text{amt. req'd for bond issue} = \$500,000$$
$$+ (0.02)(\$500,000)$$
$$= \$510,000$$

Determine the annual cost.

$$\text{annual cost} = (\text{annual yield})$$
$$\times (\text{amt. req'd for bond issue}) + A$$
$$= (0.06)(\$510,000) + (\$510,000)$$
$$\times \left(\frac{0.045}{(1+0.045)^{10} - 1}\right)$$
$$= \$72,103 \quad (\$72,100)$$

The answer is (D).

Why Other Options Are Wrong

(A) This incorrect answer is the annual cost of the bond interest only.

(B) This incorrect answer is the annual cost of the sinking fund necessary to pay off the bond face value only.

(C) This incorrect answer assumes that the 2% bond sales cost is not a financable amount but is deducted from the bond face value.

SOLUTION 100

Deficiencies that are most related to efficient use of road space are I, II, III, V, VI, and VIII.

Deficiencies I, III, and VI relate to capacity and level of service, which are measures of roadway efficiency.

Deficiencies II and V relate to managing traffic flow.

Deficiencies IV and VII relate to construction standards that could have an effect on traffic flow under severe conditions but that are not as directly related as other deficiencies listed.

Deficiency IX relates to the number of truck companies servicing businesses. The number of companies has no bearing on traffic flow; rather, it is the number of trucks themselves that is important.

The deficiencies least likely to qualify for TSM programs are IV, drainage, VII, lighting, and IX, the number of truck companies.

The answer is (B).

Why Other Options Are Wrong

(A) The answer is incorrect. Deficiency III, the peak average travel speed below 20 mph, is strongly related to traffic management and efficient use of road space.

(C) The answer is incorrect. Deficiencies VI, low level of service, and VIII, high accident rate, strongly relate to traffic management and efficient use of road space.

(D) The answer is incorrect. Deficiency V, sight distance limitations, strongly relates to traffic management and efficient use of road space.

Turn to PPI for Your Civil PE Exam Review Materials
The Most Trusted Source for Civil PE Exam Preparation
Visit www.ppi2pass.com today!

The Most Comprehensive Reference Materials

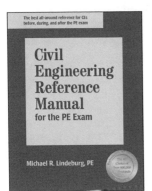

Civil Engineering Reference Manual for the PE Exam
Michael R. Lindeburg, PE

- Most widely used civil PE exam reference
- Provides the topical review and support material you need
- Thorough introduction to exam format and content
- Full Glossary for quick reference
- Quickly locate information through the complete index
- Most comprehensive exam-focused publication

Quick Reference for the Civil Engineering PE Exam
Michael R. Lindeburg, PE

- Quickly and easily access the formulas needed most often during the exam
- Drawn from the *Civil Engineering Reference Manual*
- Organized by topic
- Indexed for rapid retrieval

The Practice You Need to Succeed

Practice Problems for the Civil Engineering PE Exam
Michael R. Lindeburg, PE

- The perfect companion to the *Civil Engineering Reference Manual*
- Over 430 practice problems increase your problem-solving skills
- Multiple-choice format, just like the exam
- Coordinated with the *Civil Engineering Reference Manual* for focused preparation
- Step-by-step solutions provide immediate feedback

Civil PE Sample Examination
Michael R. Lindeburg, PE

- The most realistic practice anywhere
- Replicates the format, difficulty, and time pressure of the exam
- 40 problems covering the breadth portion
- 40 problems covering each of the depth portions
- Step-by-step solutions improve your problem-solving methods

Civil Engineering Solved Problems
Michael R. Lindeburg, PE

- Improve your problem-solving skills
- Collection of over 370 problems in scenario format
- Higher level of difficulty than problems presented on the exam
- Step-by-step solutions included for each problem

Six-Minute Solutions for Civil PE Exam Problems

5 individual books cover each of the discipline-specific topics:

Environmental	Transportation
Geotechnical	Water Resources
Structural	

- Learn to solve problems in under 6 minutes
- Improve your problem-solving speed and skills
- Perfect for the breadth and depth portions
- Discover how to avoid common mistakes
- 100 multiple-choice problems with solutions

For the latest civil PE exam news, the latest test-taker advice, the unique community of the Exam Forum, the Exam Cafe, and FAQs, go to **www.ppi2pass.com**.

Professional Publications, Inc.
www.ppi2pass.com

Turn to PPI for All of Your Exam Preparation Needs

PPI is your one stop for review manuals, practice problems, sample exams, quick references, and much more!

Visit www.ppi2pass.com to see our complete selection
of review materials for the FE and PE exams.

FE Exam Review

FE Review
FE Review Manual

EIT Review
Engineer-In-Training Reference Manual

EIT Solutions
Solutions Manual for the Engineer-In-Training Reference Manual

Sample Exams
FE/EIT Sample Examinations

Civil PE Exam Review

Reference Manual
Civil Engineering Reference Manual for the PE Exam

Practice Problems
Practice Problems for the Civil Engineering PE Exam

Sample Exam
Civil PE Sample Examination

Quick Reference
Quick Reference for the Civil Engineering PE Exam

Mechanical PE Exam Review

Reference Manual
Mechanical Engineering Reference Manual for the PE Exam

Practice Problems
Practice Problems for the Mechanical Engineering PE Exam

Sample Exam
Mechanical PE Sample Examination

Quick Reference
Quick Reference for the Mechanical Engineering PE Exam

Electrical PE Exam Review

Reference Manual
Electrical Engineering Reference Manual for the PE Exam

Practice Problems
Practice Problems for the Electrical and Computer Engineering PE Exam

Sample Exam
Electrical and Computer PE Sample Examination

Quick Reference
Quick Reference for the Electrical and Computer Engineering PE Exam

Environmental PE Exam Review

Reference Manual
Environmental Engineering Reference Manual

Practice Problems
Practice Problems for the Environmental Engineering PE Exam

Practice Exams
Environmental Engineering Practice PE Exams

Chemical PE Exam Review

Reference Manual
Chemical Engineering Reference Manual for the PE Exam

Practice Problems
Practice Problems for the Chemical Engineering PE Exam

Solved Problems
Chemical Engineering Solved Problems

Quick Reference
Quick Reference for the Chemical Engineering PE Exam

Structural PE Exam Review

Reference Manual
Structural Engineering Reference Manual for the PE Exam

Solved Problems
246 Solved Structural Engineering Problems

Get the Latest Exam News
Sign up to receive the most current news for your specific exam. Updates include notices of exam changes, useful exam tips, errata postings, and new product announcements. And, they're free!
www.ppi2pass.com/emailupdates

Order today!
Visit www.ppi2pass.com
or call 800-426-1178.

Professional Publications, Inc.
www.ppi2pass.com

Promotion Code: **EBIS**